Sidney Luxton Loney

A Treatise on Elementary Dynamics

Sidney Luxton Loney

**A Treatise on Elementary Dynamics**

ISBN/EAN: 9783337277451

Printed in Europe, USA, Canada, Australia, Japan

Cover: Foto ©berggeist007 / pixelio.de

More available books at **www.hansebooks.com**

# A TREATISE

ON

# ELEMENTARY DYNAMICS

London: C. J. CLAY AND SONS,
CAMBRIDGE UNIVERSITY PRESS WAREHOUSE,
AVE MARIA LANE.

CAMBRIDGE: DEIGHTON, BELL, AND CO.
LEIPZIG: F. A. BROCKHAUS.

# A TREATISE

## ON

# ELEMENTARY DYNAMICS

BY

S. L. LONEY, M.A.
FELLOW OF SIDNEY SUSSEX COLLEGE, CAMBRIDGE.

CAMBRIDGE:
AT THE UNIVERSITY PRESS.
1889

[*All Rights reserved.*]

Cambridge:
PRINTED BY C. J. CLAY, M.A. AND SONS,
AT THE UNIVERSITY PRESS.

# PREFACE.

In the following work I have attempted to write a fairly complete text-book on those parts of Dynamics which can be treated without the use of the Infinitesimal Calculus.

The book is intended for the use of beginners, but it is probable that most students would find it advisable, at any rate on the first reading of the subject, to confine their attention to the examples given in the text, reserving most of the questions at the ends of the chapters for a second reading. Most students would also find it advisable to omit Chap. III., the last four articles of Chap. IV., and Chap. IX., until the rest of the book has been fairly mastered.

I have ventured to alter slightly the time-honoured enunciation of Newton's Second Law of Motion.

I have used the property of the hodograph in the chapter on normal acceleration and in dealing with the motion of a particle in a conic section with an acceleration directed towards the focus.

L. D.

In treating cycloidal and pendulum motion I have first introduced the student to the conception of simple harmonic motion and thus have avoided much of the dreary work usually to be found in this portion of the subject.

I must express my gratitude to Mr C. SMITH, M.A., Fellow and Tutor of Sidney Sussex College, Cambridge, for much kind criticism and encouragement; also to Mr H. C. ROBSON, M.A., of Sidney Sussex College, and other friends.

I have spared no pains to ensure the accuracy of the answers to the examples, and hope that not many serious errors will be found. I could not, however, burden my friends with the large amount of drudgery involved in their verification and so was obliged to rely on my own unaided resources.

For any corrections, or suggestions for improvement, I shall be very grateful.

S. L. LONEY.

LONSDALE ROAD, BARNES. S. W.
*July* 17, 1889.

# CONTENTS.

## CHAPTER I.

### UNIFORM AND UNIFORMLY ACCELERATED MOTION.

|  | PAGE |
|---|---|
| Dynamics. Definition and Subdivisions | 1 |
| Speed | 2 |
| Velocity | 4 |
| Parallelogram of **Velocities** | 7 |
| Component and resultant velocities | 8 |
| Triangle and Parallelopiped of Velocities | 10 |
| Examples | 12 |
| Change of velocity **and acceleration** | 15 |
| Parallelogram of Accelerations | 16 |
| Relative Velocity. **Examples** | 18 |
| Angular and **Areal Velocity. Examples** | 24 |
| Formulae to determine the **motion when the acceleration and initial** velocity are given. **Examples** | 28 |
| Acceleration of falling **bodies.** Morin's experiment | 33 |
| Vertical motion under **gravity.** Examples | 34 |
| Motion on a smooth inclined plane. **Examples** | 38 |
| Time down chords of a vertical circle | 40 |
| Lines of quickest descent | 41 |
| Graphic Methods. **Velocity-Time Curve.** Acceleration-Time Curve | 46 |
| Examples on Chapter I. | 50 |

## CHAPTER II.
### THE LAWS OF MOTION.

| | PAGE |
|---|---|
| Definitions | 56 |
| Enunciation of the Laws of Motion | 60 |
| The relation $P = mf$ | 62 |
| Definition of the unit of force | 63 |
| Absolute and gravitation units | 64 |
| The weight of a body is proportional to its mass | 64 |
| Distinction between mass and weight | 65 |
| Weighing by scales and a spring-balance | 66 |
| Examples | 67 |
| Physical independence of forces | 69 |
| Parallelogram of Forces | 70 |
| Examples of the Third Law of Motion | 71 |
| Motion of two particles connected by a string placed over a smooth pulley | 72 |
| Atwood's machine | 73 |
| Examples | 75 |
| Impulse and impulsive forces | 81 |
| Impact of two bodies | 84 |
| Motion of a shot and gun | 85 |
| Work and Power | 86 |
| Examples | 89 |
| Kinetic and Potential Energy | 92 |
| Conservation of Energy | 93 |
| Graphic Methods. Force-Space Curve | 95 |
| Motion of the centre of inertia of a system of bodies | 96 |
| Conservation of Momentum | 100 |
| Miscellaneous Examples | 101 |
| Examples on Chapter II. | 103 |

## CHAPTER III.
### THE LAWS OF MOTION (*continued*). MISCELLANEOUS EXAMPLES.

| | |
|---|---|
| Motion of two pulleys, each carrying masses, connected by a string passing over a fixed pulley | 114 |
| Motion of a wheel-and-axle | 116 |

CONTENTS. ix

|   | PAGE |
|---|---|
| Motion of a movable wedge with masses upon its faces | 117 |
| Ratio of velocity of a cannon-ball when the cannon is free to move to the velocity when the cannon is fixed | 119 |
| Perforation of a plate by a shot | 120 |
| Pile-driving | 121 |
| Man leaping from a movable platform | 122 |
| Motion of a heavy elastic ring upon a smooth cone | 123 |
| Starting of a goods train with the couplings slack | 123 |
| Motion of heavy strings | 125 |
| Transmission of power by belts and shafting | 128 |
| Examples on Chapter III. | 129 |

## CHAPTER IV.

### UNITS AND DIMENSIONS.

|   | |
|---|---|
| Fundamental and derived units | 143 |
| The unit of velocity varies directly as the unit of length, and inversely as the unit of time | 145 |
| The unit of acceleration varies directly as the unit of length, and inversely as the square of the unit of time | 146 |
| Definition of dimensions | 148 |
| Dimensions of Velocity, Force, Work, etc. | 148 |
| Examples | 151 |
| Verification of formulae by means of counting the dimensions | 156 |
| Attraction units | 158 |
| Astronomical unit of mass | 161 |
| Table of dimensions and values of fundamental quantities | 162 |
| Examples on Chapter IV. | 164 |

## CHAPTER V.

### PROJECTILES.

|   | |
|---|---|
| Definitions and properties of a parabola | 167 |
| The path of a projectile is a parabola | 170 |
| The velocity at any point is that due to a fall from the directrix | 171 |
| Greatest height and range on a horizontal plane. Maximum range | 173 |

CONTENTS.

|  | PAGE |
|---|---|
| Two directions of projection for a given range | 174 |
| Latus-rectum of the path | 175 |
| Velocity and direction of motion at any time | 175 |
| Focus and directrix of path | 176 |
| Equation to the path | 177 |
| Examples | 177 |
| Range on an inclined plane. Maximum range | 181 |
| Motion upon an inclined plane | 183 |
| Examples | 184 |
| Geometrical construction for the path | 185 |
| Envelope of paths | 186 |
| Geometrical construction for maximum range | 188 |
| Miscellaneous Examples | 189 |
| Examples on Chapter V. | 191 |

## CHAPTER VI.

### COLLISION OF ELASTIC BODIES.

|  |  |
|---|---|
| Elasticity | 202 |
| Newton's Experimental Law | 203 |
| Impact on a smooth fixed plane | 205 |
| Direct impact of two spheres | 207 |
| Oblique impact of two spheres | 209 |
| Examples | 212 |
| Loss of Kinetic Energy by impact | 217 |
| Action between two elastic bodies during their collision | 219 |
| Impact of a particle on a rough plane | 220 |
| Miscellaneous Examples | 222 |
| Examples on Chapter VI. | 227 |

## CHAPTER VII.

### THE HODOGRAPH AND NORMAL ACCELERATIONS.

|  |  |
|---|---|
| Definition of the hodograph | 241 |
| The velocity in the hodograph represents the acceleration in the path | 242 |

CONTENTS. xi

PAGE

Normal acceleration of a point moving in a circle with uniform
 speed . . . . . . . . . . . 243
Normal and tangential accelerations in any curve . . . . 245
" Centrifugal Force " . . . . . . . . . . 247
Examples . . . . . . . . . . . 248
The conical pendulum . . . . . . . . . 250
Railway carriage on a curved line . . . . . . 252
Rotating sphere . . . . . . . . . . 254
Revolving string . . . . . . . . . . 254
Examples . . . . . . . . . . . 255
On a smooth curve, under gravity only, the change of velocity is
 that due to the vertical distance through which the particle
 moves . . . . . . . . . . . . 257
Newton's Experimental Law . . . . . . . . 260
Motion on the outside of a vertical circle . . . . . 261
Motion in a vertical circle . . . . . . . . 262
Effect of the rotation of the earth upon the apparent weight of a
 particle . . . . . . . . . . . . 265
Examples on Chapter VII. . . . . . . . . 268

# CHAPTER VIII.

## SIMPLE HARMONIC MOTION. CYCLOIDAL AND PENDULUM MOTIONS.

Definition and investigation of simple harmonic motion . . . 277
Extension to motion in a curve . . . . . . . 281
Examples . . . . . . . . . . . 282
Definition and properties of a cycloid . . . . . . 284
Isochronism of a cycloid . . . . . . . . . 287
Simple pendulum . . . . . . . . . . 288
Small oscillations in a vertical circle . . . . . . 289
Seconds pendulum . . . . . . . . . . 290
Simple equivalent pendulum . . . . . . . . 291
Effect on the time of oscillation of small changes in the value of
 "$g$" and of small changes in the length of the pendulum . 292
Finding "$g$" by means of the pendulum . . . . . 294
Verification of law of gravity by means of the moon's motion . 295
Examples on Chapter VIII. . . . . . . . . 295

## CHAPTER IX.

### MOTION OF A PARTICLE ABOUT A FIXED CENTRE OF FORCE.

|  | PAGE |
|---|---|
| Definition of the moment of a velocity | 300 |
| The moment of the velocity of a particle about a fixed centre, to which its acceleration is always directed, is constant | 301 |
| Motion in an ellipse about the centre | 302 |
| Motion in a conic section about the focus | 305 |
| Kepler's Laws | 308 |
| Effect of disturbing forces on the path of a particle | 309 |
| Effect of a change in the absolute value of the acceleration | 310 |
| Motion in a straight line with an acceleration varying inversely as the square of the distance | 310 |
| Motion of a projectile, variations of gravity being considered | 311 |
| Miscellaneous Examples | 314 |
| Examples on Chapter IX. | 315 |
| ANSWERS TO THE EXAMPLES | 321 |

# CHAPTER I.

### UNIFORM AND UNIFORMLY ACCELERATED MOTION.

1. **Dynamics** is the science which treats of the action of force on bodies.

When a body is acted on by one or more forces, their effect is either (1) to compel rest or prevent motion, or (2) to produce or change motion.

Dynamics is therefore conveniently divided into two portions, **Statics** and **Kinetics.**

In Statics the subject of the equilibrium, or balancing, of forces is considered; in Kinetics is discussed the action of forces in producing, or changing, motion.

The present book treats only of Kinetics.

The branch of Pure Mathematics which deals with the geometry of the motion of bodies, without any reference to the forces which produce or change the motion, is called **Kinematics.** It is a necessary preliminary to Dynamics.

The present chapter will be devoted to Kinematics; in the next chapter we shall enter on Dynamics.

2. The position of a point in a plane is generally determined in one of two ways. We may choose a point

$O$ fixed in the plane, and a fixed straight line $Ox$ through $O$; then the position of a point $P$ is known, when the distance $OP$ and the angle $xOP$ are known.

Or again we may choose a fixed point $O$ and two fixed straight lines $Ox$, $Oy$ drawn from $O$; from a point $P$ draw $PM$, $PN$ parallel to $Oy$, $Ox$ respectively to meet $Ox$, $Oy$ in $M$, $N$. The position of $P$ is now known if the distances $MP$, $NP$ are known.

Here we observe that to fix the position of a point in a plane we must have two quantities given, viz. either a length and a direction, or two lengths in two fixed directions.

3. *Change of position.* If at any instant the position of a moving point is $P$, and at any subsequent time it is $Q$, then $PQ$ is the change of its position in that time.

A point is said to be in motion when it changes its position.

The path of a moving point is the curve drawn through all the successive positions of the point.

4. **Speed. Def.** *The speed of a moving point is the rate at which it describes its path.*

A point is said to be moving with uniform speed when it moves through equal lengths in equal times, however small these times may be.

Suppose a train describes 30 miles in each of several consecutive hours. We are not justified in saying that its speed is uniform unless we know that it describes half a mile in *each* minute, 44 feet in *each* second, one-millionth of 30 miles in *each* one-millionth of an hour, and so on.

When uniform, the speed of a point is measured by the distance passed over by it in a unit of time; when variable, by the distance which would be passed over by the point in a unit of time, if it continued to move during that unit of time with the speed it has at the instant under consideration.

By saying that a train is moving with a speed of 40 miles an hour, we do not mean that it has gone 40 miles in the last hour, or that it will go 40 miles in the next hour; but that if its speed remained constant for one hour then it would describe 40 miles in that hour.

5. The definition of the measure of the speed of a point at any instant may also be stated as follows:

Let $\sigma$ be the length of the portion of the path passed over by the moving point in the time $\tau$ following the instant under consideration; then the limiting value of the ratio $\dfrac{\sigma}{\tau}$, as the time $\tau$ is taken indefinitely small, is the measure of the speed at that instant.

6. The units of length and time usually employed in England are a foot and a second.

In scientific measurements the unit of length usually employed is the centimetre, which is the one hundredth part of a metre.

One metre = 39·37 inches approximately. A decimetre is $\frac{1}{10}$th, and a millimetre $\frac{1}{1000}$th of a metre.

7. The unit of speed is the speed of a point which moves over a unit of length in a unit of time. Hence the unit of speed depends on these two units, and if either, or both of them, be altered the unit of speed will also, in general, be altered.

8. If a point is moving with speed $u$, then in each unit of time the point moves over $u$ units of length.

∴ in $t$ units of time the point passes over $ut$ units of length.

Hence the distance $s$ passed over by a point which moves with speed $u$ for time $t$ is given by $s = ut$.

**Ex. 1.** Find the speed of the centre of the earth in metres per second, assuming that it describes a circle of radius 93,000,000 miles in 365 days.

**Ex. 2.** Find the speed of light in miles per second, if it takes 8 minutes to describe the distance from the sun to the earth.

**Ex. 3.** Find the speed, in metres per second, arising from the rotation of the earth, of a point in latitude 60°.

**9. Displacement.** The displacement of a moving point is its change of position. To know the displacement of a moving point we must know both the length of the line joining the two positions of the point, and also the direction. Hence the displacement of a point involves both magnitude and direction.

**Ex. 1.** A man walks 3 miles due east, and then 4 miles due north; find his displacement.

*Ans.* 5 miles at an angle $\tan^{-1} \frac{4}{3}$ north of east.

**Ex. 2.** A ship sails 1 mile due south and then $\sqrt{2}$ miles south-west; shew that its displacement is $\sqrt{5}$ miles in a direction $\tan^{-1} \frac{1}{2}$ west of south.

**10. Velocity. Def.** *The velocity of a moving point is the rate of its displacement.*

A velocity therefore possesses both magnitude and direction.

A point is said to be moving with uniform velocity, when it is moving in a constant direction, and passes over equal lengths in equal times, however small these times may be.

When uniform the velocity of a moving point is measured by its displacement per unit of time; when

VELOCITY. 5

variable, it is measured, at any instant, by the displacement that the moving point would have in a unit of time, if it moved during that unit of time with the velocity which it has at the instant under consideration.

11. The definition of the measure of the velocity of a point at a given instant may also be stated as follows:

Let $d$ be the displacement of a moving point in the time $\tau$ following the instant under consideration; then the limiting value of $\dfrac{d}{\tau}$, as the time $\tau$ is taken indefinitely small, is the measure of the velocity at that instant.

12. It will be noted that when the moving point is moving in a straight line, the velocity is the same as the speed. The word velocity is often used to express what we in Art. 4 call the speed of the moving point. It seems however desirable, at any rate for the beginner, to carefully distinguish between the two.

If the motion be not in a straight line the velocity is not the same as the speed. For example, suppose a point to be describing a circle uniformly so that it passes over equal lengths of the arc in equal times however small; its direction of motion (viz. the tangent to the circle) is different at different points of the circumference; hence the *velocity* of the point (strictly so called) is variable, whilst its *speed* is constant.

13. The unit of velocity is the velocity of a point which undergoes a displacement equal to a unit of length in a unit of time.

When we say that a moving point has velocity $v$, we mean that it possesses $v$ units of velocity, i.e. that it would undergo a displacement, equal to $v$ units of length, in the unit of time.

If the velocity of a moving point in one direction be denoted by $v$, an equal velocity in an opposite direction is necessarily denoted by $-v$.

14. Since the velocity of a point is known when its direction and magnitude are both known, we can conveniently represent the velocity of a moving point by a straight line $AB$; thus when we say that the velocities of two moving points are represented in magnitude and direction by the straight lines $AB$, $CD$, we mean that they move in directions parallel to the lines drawn from $A$ to $B$, and $C$ to $D$ respectively, and with velocities which are proportional to the lengths $AB$ and $CD$.

15. A body may have simultaneously velocities in two or more different directions. One of the simplest examples of this is when a person walks on the deck of a ship in motion from one point of the deck to another. He has a motion with the ship, and one along the surface of the ship, and his motion in space is clearly different from what it would have been had either the ship remained at rest, or had the man stayed at his original position on the deck.

Again consider the case of a ship steaming with its bow pointing in a constant direction, say due north, whilst a current carries it in a different direction, say south-east, and suppose a sailor is climbing a vertical mast of the ship. The actual change of position and the velocity of the sailor clearly depend on three quantities, viz. the rate and direction of the ship's sailing, the rate and direction of the current, and the rate at which he climbs the mast. His actual velocity is said to be "compounded" of these three velocities.

In the following article we shew how to find the velocity which is equivalent to two velocities given in magnitude and direction.

16. **Theorem. Parallelogram of Velocities.** *If a moving point possess simultaneously* velocities *which are represented in magnitude and direction by the two sides of a parallelogram drawn from a point, they are equivalent to a velocity which is represented in magnitude and direction by the diagonal of the parallelogram passing through the point.*

Let the two simultaneous velocities be represented by the lines $AB$, $AC$ and let their magnitudes be $u$, $v$.

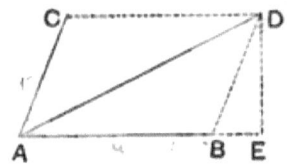

Complete the parallelogram $BACD$.

Then we may imagine the motion of the point to be along the line $AB$ with the velocity $u$, whilst the page of this book moves parallel to $AC$ with velocity $v$. In the unit of time the moving point will have moved through a distance $AB$ along the line $AB$, and the line $AB$ will have in the same time moved into the position $CD$ so that the moving point will now be at $D$.

Now, since the two coexistent velocities are constant in magnitude and direction, the velocity of the point from $A$ to $D$ must be also constant in magnitude and direction; hence $AD$ is the path described by the moving point in the unit of time.

Hence $AD$ represents in magnitude and direction the velocity which is equivalent to the velocities represented by $AB$ and $AC$.

17. **Def.** *The velocity which is equivalent to two or more velocities is called their* **resultant**, *and these velocities are called the* **components** *of this resultant.*

The resultant of two velocities $u$, $v$ in directions which are inclined to one another at a given angle $\alpha$ may be easily obtained.

For in Fig. Art. 16, let $AB$, $AC$ represent the velocities $u$, $v$, so that $\angle BAC = \alpha$.

Then $AD^2 = AB^2 + BD^2 - 2AB \cdot BD \cos ABD$.

Hence if we represent the resultant velocity $AD$ by $w$, we have

$$w^2 = u^2 + v^2 + 2uv \cos \alpha, \quad \text{since } \angle ABD = \pi - \alpha.$$

Also $\dfrac{\sin BAD}{\sin ABD} = \dfrac{BD}{AB}$;

$\therefore \dfrac{\sin \theta}{\sin (\alpha - \theta)} = \dfrac{v}{u}$, where $\angle BAD = \theta$;

$\therefore \tan \theta = \dfrac{v \sin \alpha}{u + v \cos \alpha}$.

*Hence the resultant of two velocities* $u$, $v$ *inclined to one another at an angle* $\alpha$, *is a velocity* $\sqrt{u^2 + v^2 + 2uv \cos \alpha}$ *inclined at an angle* $\tan^{-1} \dfrac{v \sin \alpha}{u + v \cos \alpha}$ *to the direction of the velocity* $u$.

The direction of the resultant velocity may also be obtained as follows; draw $DE$ perpendicular to $AB$ to meet it, produced if necessary, in $E$; we then have

$$\tan DAB = \frac{DE}{AE} = \frac{BD \sin EBD}{AB + BD \cos EBD} = \frac{v \sin \alpha}{u + v \cos \alpha}.$$

18. A velocity can be resolved into two component velocities in an infinite number of ways. For an infinite number of parallelograms can be described having a line

$AD$ as diagonal; and if $ACDB$ be any one of these the velocity $AD$ is equivalent to the two component velocities $AB$ and $AC$.

The most important case is when a velocity is to be resolved into two velocities in two directions *at right angles,* one of these directions being given. When we speak of the *component of a velocity in a given direction* it is to be understood that the other direction in which the given velocity is to be resolved is perpendicular to this given direction.

Thus suppose we wish to resolve a velocity $u$ represented by $AD$ into two components at right angles to one another, one of these components being along a line $AB$ making an angle $\theta$ with $AD$.

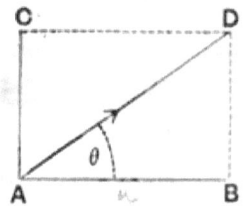

Draw $DB$ perpendicular to $AB$, and complete the rectangle $ABDC$.

Then the velocity $AD$ is equivalent to the two component velocities $AB, AC$.

Also $AB = AD \cos \theta = u \cos \theta$,
and $AC = AD \sin \theta = u \sin \theta$.

We thus have the following important

**Theorem.** *A velocity* u *is equivalent to a velocity* u cos θ *along a direction making an angle θ with its own direction together with a velocity* u sin θ *perpendicular to the direction of the first component.*

**Ex. 1.** A man is walking in a north-easterly direction with a velocity of 4 miles per hour; find the components of his velocity in directions due north and due east respectively.

**Ex. 2.** A point is moving in a straight line with a velocity of 10 feet per second; find the component of its velocity in a direction inclined at an angle of 30° to its direction of motion.

**19.** *Components of a velocity in two given directions.*

If we wish to find the components of a velocity $u$ in two given directions making angles $\alpha$, $\beta$ with it, we proceed as follows.

Let $AD$ represent $u$ in magnitude and direction. Draw $AB$, $AC$ making angles $\alpha$, $\beta$ with it, and through $D$ draw parallels to complete the parallelogram $ABDC$ as in Fig. Art. 16. Then we have

$$\frac{AB}{\sin \beta} = \frac{BD}{\sin \alpha} = \frac{AD}{\sin(\alpha+\beta)}.$$

$$\therefore AB = AD \cdot \sin\beta / \sin(\alpha+\beta),$$
$$AC = AD \cdot \sin\alpha / \sin(\alpha+\beta).$$

Hence the component velocities in these two directions are

$$u\sin\beta/\sin(\alpha+\beta), \text{ and } u\sin\alpha/\sin(\alpha+\beta).$$

**20. Triangle of Velocities.** *If a moving point possess simultaneously velocities represented by the two sides* AB, BC *of a triangle taken in order, they are equivalent to a velocity represented by* AC.

For completing the parallelogram $ABCD$ the lines $AB$, $BC$ represent the same velocities as $AB$, $AD$ and hence have as their resultant the velocity represented by $AC$.

*Cor.* 1. If there be simultaneously impressed on a point three velocities represented by the sides of a triangle taken in order the point is at rest.

*Cor.* 2. If a moving point possesses velocities represented by $\lambda \cdot OA$ and $\mu \cdot OB$, they are equivalent to a velocity $(\lambda+\mu) \cdot OG$ where $G$ is a point on $AB$ such that

$$\lambda \cdot AG = \mu \cdot BG.$$

For by the triangle of velocities a velocity $\lambda \cdot OA$ is equivalent to velocities $\lambda \cdot OG$ and $\lambda \cdot GA$; also the velocity $\mu \cdot OB$ is equivalent to $\mu \cdot OG$ and $\mu \cdot GB$; but the velocities $\lambda \cdot GA$ and $\mu \cdot GB$ destroy one another; hence the resultant velocity is $(\lambda+\mu) \cdot OG$.

**21. Parallelopiped of Velocities.** By a proof similar to that for the parallelogram of velocities it may be shewn that the resultant of three velocities represented by the three edges of a parallelopiped meeting in a point, is a velocity represented by the diagonal of the parallelopiped passing through that angular point. Conversely a velocity may be resolved into three others.

Similarly, as in Art. 18, it follows that a velocity $u$ may be resolved into velocities $u \cos \alpha$, $u \cos \beta$, $u \cos \gamma$ along three directions in space mutually at right angles, where $\alpha$, $\beta$, $\gamma$ are the angles that the direction of $u$ makes with these directions.

22. If a moving point possess simultaneously velocities represented by the sides $AB$, $BC$, $CD$,...$KL$ of a polygon, (whether the sides of the polygon are or are not in one plane) the resultant velocity is represented by $AL$.

For, by Art. 20, the velocities $AB$, $BC$ are represented by $AC$; and again the velocities $AC$, $CD$ by $AD$ and so on; so that the final velocity is represented by $AL$.

*Cor.* If the point $L$ coincides with $A$ (so that the polygon is a closed figure) the resultant velocity vanishes, and the point is at rest.

23. When a point possesses simultaneously velocities in several different directions in the same plane their resultant may be found by resolving the velocities along two fixed directions at right angles, and then compounding the resultant velocities.

Suppose a point possesses velocities $u$, $v$, $w$... in directions inclined at angles $\alpha$, $\beta$, $\gamma$,... to a fixed line $Ox$, and let $Oy$ be perpendicular to $Ox$. The components of $u$

along $Ox$, $Oy$ are respectively $u\cos\alpha$, $u\sin\alpha$; the components of $v$ are $v\cos\beta$, $v\sin\beta$; and so for the others.

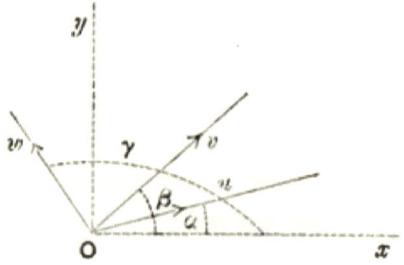

Hence the velocities are equivalent to

$u\cos\alpha + v\cos\beta + w\cos\gamma$ ......parallel to $Ox$,

and $u\sin\alpha + v\sin\beta + w\sin\gamma$ ......parallel to $Oy$.

If their resultant be a velocity $V$ at an angle $\theta$ to $Ox$ we must have

$V\cos\theta = u\cos\alpha + v\cos\beta + w\cos\gamma + ......$,

and $V\sin\theta = u\sin\alpha + v\sin\beta + w\sin\gamma + ......$

Hence, by squaring and adding,

$V^2 = (u\cos\alpha + v\cos\beta + ...)^2 + (u\sin\alpha + v\sin\beta + ...)^2$;

and, by division, $\tan\theta = \dfrac{u\sin\alpha + v\sin\beta + ...}{u\cos\alpha + v\cos\beta + ...}$.

These two equations give $V$ and $\theta$.

### EXAMPLES.

1. *A vessel steams with its bow pointed due north with a velocity of 15 miles an hour, and is carried by a current which flows in a south-easterly direction at the rate of $3\sqrt{2}$ miles per hour. Find its distance and bearing from the point from which it started at the end of an hour.*

The ship has two velocities one 15 miles per hour northwards and another $3\sqrt{2}$ miles per hour south-east.

## EXAMPLES.

Now the latter is equivalent to

$3\sqrt{2}\cos 45°$ or 3 miles per hour eastward,

and $\quad 3\sqrt{2}\sin 45°$ or 3 miles per hour southward.

∴ the total velocity of the ship is 12 miles per hour northwards and 3 miles per hour eastward.

Hence its resultant velocity is $\sqrt{12^2+3^2}$ or $\sqrt{153}$ miles per hour in a direction inclined at an angle $\tan^{-1}\frac{1}{4}$ to the north.

**2.** *A point possesses simultaneously velocities whose measures are* 4, 3, 2, 1; **the** *angle between the first and second is* 30°, *between the second and third* 90°, *and between the* **third and** *fourth* 120°; *find their resultant.*

**Take** $Ox$ along the direction **of the first velocity and** $Oy$ perpendicular to it.

The angles the velocities make with $Ox$ are respectively 0°, 30°, 120°, and 240°.

Hence, if $V$ be the resultant **velocity inclined at an angle** $\theta$ **to** $Ox$, we have

$V\cos\theta = 4 + 3\cos 30° + 2\cos 120° + 1 . \cos 240°;$

and $\quad V\sin\theta = \quad 3\sin 30° + 2\sin 120° + 1 . \sin 240°.$

Whence we have

$$V\cos\theta = \frac{5+3\sqrt{3}}{2}, \quad V\sin\theta = \frac{3+\sqrt{3}}{2}.$$

Thence $\quad V^2 = 16 + 9\sqrt{3},$

and $\quad \tan\theta = \dfrac{3+\sqrt{3}}{5+3\sqrt{3}} = 2\sqrt{3} - 3.$

Hence the resultant is a velocity $\sqrt{16+9\sqrt{3}}$ inclined at an angle $\tan^{-1}(2\sqrt{3}-3)$ to the direction of the first velocity.

**3.** The velocity of a ship is $8\frac{2}{11}$ miles per hour, and a ball is bowled across the ship perpendicular **to the** direction of the ship with a velocity of 3 yards per **second; describe the path of the ball in space** and shew that it passes over 45 feet **in 3 seconds.**

**4.** A boat is rowed with a velocity of 6 miles per hour straight across a river which flows at the rate of 2 miles per hour. If its breadth **be** 300 feet, find how far down the river the boat will reach the opposite bank, below the point at which it was originally directed.

*Ans.* **100 feet.**

# EXAMPLES.

5. A ship is steaming in a direction due north across a current running due west. At the end of one hour it is found that the ship has made $8\sqrt{3}$ miles in a direction 30° west of north. Find the velocity of the current, and the rate at which the ship is steaming.

*Ans.* $4\sqrt{3}$ miles per hour; 12 miles per hour.

6. Find the components of a velocity $u$ resolved along two lines inclined at angles of 30° and 45° respectively to its direction.

*Ans.* $(\sqrt{3}-1)u$; $(\sqrt{6}-\sqrt{2})\dfrac{u}{2}$.

7. A point which possesses velocities represented by 7, 8, 13 is at rest; find the angle between the directions of the two lesser velocities.

*Ans.* 60°.

8. A point possesses velocities represented by 3, 19, 9 inclined at angles of 120° to one another; find their resultant.

*Ans.* 14 at an $\angle \sin^{-1} \dfrac{3\sqrt{3}}{14}$ with the greatest velocity.

9. A point possesses simultaneously velocities represented by $u$, $2u$, $3\sqrt{3}u$ and $4u$; the angles between the first and second, the second and third, and the third and fourth, are respectively 60°, 90° and 150°; shew that the resultant is $u$ in a direction inclined at an angle of 120° to that of the first velocity.

10. A tram-car is moving along a road at the rate of 8 miles per hour; in what direction must a body be projected from it with a velocity of 16 feet per second, so that its resultant motion may be at right angles to the tram-car?

*Ans.* The direction must make an angle $\cos^{-1}(-\frac{11}{15})$ with the direction of the car's motion.

11. A ship is sailing north at the rate of 4 feet per second; the current is taking it east at the rate of 3 feet per second, and a sailor is climbing a vertical pole at the rate of 2 feet per second; find the velocity and direction of the sailor in space.

12. A point has equal velocities in two given directions; if one of these velocities be halved, the angle the resultant makes with the other is halved also. Shew that the angle between the velocities is 120°.

13. A point possesses velocities represented in magnitude and direction by the lines joining any point on a circle to the ends of a diameter; shew that their resultant is represented by the diameter through the point.

**14.** Two steamers $X$, $Y$ are respectively at two points $A$, $B$, 5 miles apart. $X$ steams away with a uniform velocity of 10 miles per hour in a direction making an angle of 60° with $AB$. Find in what direction $Y$ must start at the same moment, if it steams with a uniform velocity of $10\sqrt{3}$ miles per hour, in order that it may just come into collision with $X$; find also at what angle it will strike $X$.

*Ans.* At an angle of 150° with $AB$ produced; it will strike $X$ at right angles.

**24. Change of Velocity.** Suppose a point at any instant to be moving with a velocity represented by $OA$ and that at some subsequent time its velocity is represented by $OB$.

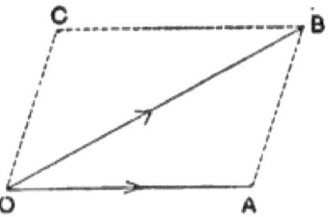

Join $AB$ and complete the parallelogram $OABC$.

Then velocities represented by $OA$, $OC$ are equivalent to the velocity $OB$. Hence the velocity $OC$ is the velocity which must be compounded with $OA$ to produce the velocity $OB$. The velocity $OC$ is therefore the change of velocity in the given time.

Thus the change of velocity is not the difference in magnitude between the magnitudes of the two velocities, but is that velocity which compounded with the original velocity gives the final velocity. In the following definition the expression "change of velocity" is used in this meaning.

**25. Acceleration. Def.** *The acceleration of a moving point is the rate of change of its velocity.*

The acceleration is uniform when equal changes of velocity take place in equal intervals of time, however small these intervals may be.

When uniform, the acceleration is measured by the change in the velocity in a unit of time; when variable, it is measured at any instant by what would be the change of the velocity in a unit of time, if during that time the acceleration continued the same as at the instant under consideration.

26. The definition of the measure of the acceleration at any instant may also be stated as follows:

Let $v$ be the velocity which must be compounded with the velocity at the given instant, to produce the velocity which the point possesses at the time $\tau$ following the given instant; then the measure of the acceleration at the given instant is the limit, when $\tau$ is made indefinitely small, of the quantity $\dfrac{v}{\tau}$.

27. The unit of acceleration is the acceleration of a point which moves so that its velocity is changed by the unit of velocity in each unit of time.

Hence a point is moving with $n$ units of acceleration when its velocity is changed by $n$ units of velocity in each unit of time.

Ex. 1. A point is moving eastwards with a velocity of 20 feet per second, and one hour afterwards it is moving north-east with the same velocity; find the change of velocity, and the measure of the acceleration, supposing the latter to be uniform.

Ans. $20\sqrt{2-\sqrt{2}}$ ft. per second in a direction N.N.W.

Ex. 2. A point is describing a circle of radius 7 yards in 11 seconds, starting from the end of a fixed diameter, and moving with uniform speed; find the change in its velocity after it has described one-sixth of the circumference.

Ans. 12 ft.-sec. units at an angle of 120° with the initial direction of motion.

28. **Theorem. Parallelogram of Accelerations.** *If a moving point have simultaneously two accelerations*

## PARALLELOGRAM OF ACCELERATIONS.

*represented in magnitude and direction by two sides of a parallelogram drawn from a point, they are equivalent to an acceleration represented by the diagonal of the parallelogram passing through that angular point.*

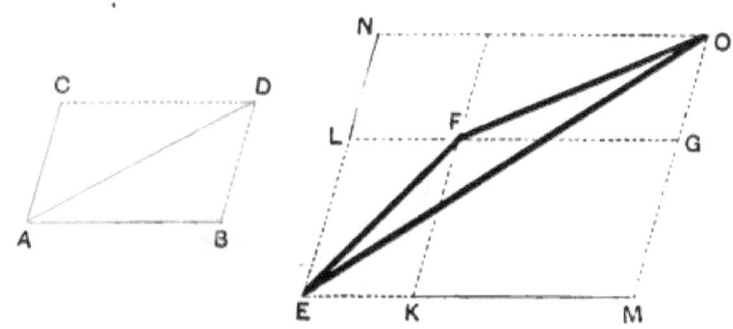

Let the accelerations be represented by the sides $AB$, $AC$ of the parallelogram $ABDC$, i.e. let $AB$, $AC$ represent the velocities added to the velocity of the point in a unit of time. On the same scale let $EF$ represent the velocity the particle has at any instant. Draw the parallelogram $EKFL$ having its sides parallel to $AB$ and $AC$; produce $EK$ to $M$, and $EL$ to $N$, so that $KM$, $LN$ are equal to $AB$, $AC$ respectively. Complete the parallelograms as in the above figure.

Then the velocity $EF$ is equivalent to velocities $EK$, $EL$. But in the unit of time the velocities $KM$, $LN$ are the changes of velocity.

∴ at the end of a unit of time the component velocities are equivalent to $EM$ and $EN$ which are equivalent to $EO$, and this latter velocity is equivalent to velocities $EF$, $FO$. (Art. 20.)

∴ in the unit of time $FO$ is the change of velocity of the moving point, i.e. $FO$ is the resultant acceleration of the point.

But $FO$ is equal and parallel to $AD$.

∴ $AD$ represents the acceleration which is equivalent to the accelerations $AB$, $AC$, i.e. is the resultant of the accelerations $AB$, $AC$.

29. It follows from the preceding article that accelerations are compounded, and resolved, in the same manner as velocities are, and propositions similar to those of Arts. 17—23 will be true if instead of the word "velocity" we substitute "acceleration".

30. **Average Speed and Velocity.** The average speed of a point in a given period of time is the same as the speed of a moving point which moves uniformly, and describes the same distance as the given point. Thus the average speed of a moving point in a given period of time is the whole distance described by the point in the given time divided by the whole time. Thus the average speed of an athlete who runs 100 yards in $10\frac{2}{5}$ seconds is $100 \div 10\frac{2}{5}$ or $9\frac{8}{13}$ yards per second.

The average velocity of a given point in any given direction (strictly so called) is the whole displacement in the given direction in the given time divided by the given time.

31. **Relative Velocity. Def.** *When the distance between two points is altering, either in direction or in magnitude or in both, then either point is said to have a velocity relative to the other, and the relative velocity of one point with respect to a second is that velocity which, compounded with the velocity of the second, gives the velocity of the first point.*

Consider the case of two trains moving on parallel rails in the same direction with equal velocities and let $A$, $B$

be two points, one on each train; a person at one of them, $A$ say, would, if he kept his attention fixed on $B$ and if he were unconscious of his own motion, consider $B$ to be at rest. The line $AB$ would remain constant in magnitude and direction and the velocity of $B$ relative to $A$ would be zero.

Next let the first train be moving at the rate of 20 miles per hour and let the second train $B$ be moving at the rate of 25 miles per hour. In this case the line joining $A$ to $B$ would be increasing at the rate of 5 miles per hour, and this would be the velocity of $B$ relative to $A$.

Thirdly let the second train be moving with a velocity of 25 miles per hour in the opposite direction to that of the first; the line joining $AB$ would now be increasing at the rate of 45 miles per hour in a direction opposite to that of $A$'s motion, and the relative velocity of $B$ with respect to $A$ would be $-45$ miles per hour.

In each of these cases it will be noticed that the relative velocity of the second train with respect to the first is obtained by compounding with its own velocity a velocity equal and opposite to that of the first.

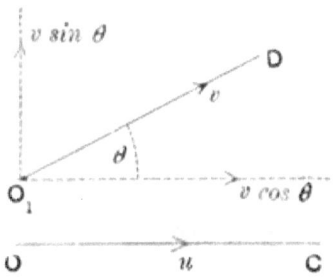

Lastly let the first train be moving along the line $OC$ with velocity $u$ whilst the second train is moving with velocity $v$ along a line $O_1D$ inclined at an angle $\theta$ to $OC$.

Resolve the velocity $v$ into two components viz. $v \cos \theta$ parallel to $OC$ and $v \sin \theta$ in the perpendicular direction.

As before, the velocity of $B$ relative to $A$, parallel to $OC$, would be $v \cos \theta - u$; also, since the point $A$ has no velocity perpendicular to $OC$, the velocity of $B$ relative to $A$ in that direction is $v \sin \theta$.

Hence the velocity of $B$ relative to $A$ would consist of two components viz. $v \cos \theta - u$ parallel to $OC$ and $v \sin \theta$ perpendicular to $OC$. These two components are equivalent to the original velocity $v$ of the train $B$ together with a velocity equal and opposite to that of $A$.

Hence we have the following important result;

*If a set of points* A, B, C, D,... *are in motion, their relative velocities with regard to any one of them,* A *say, are obtained by compounding with their velocities a velocity equal and opposite to that of* A.

32. Rest and motion are relative terms; we do not know what absolute motion is; all motion that we become acquainted with is relative.

For example when we say that a train is travelling northward at the rate of 40 miles an hour we mean that that is its velocity relative to the earth. Besides this motion along the surface it partakes with the rest of the earth in the diurnal motion about the axis of the earth; it also moves with the earth round the sun; and in addition has, in common with the whole solar system, any velocity that that system may have.

33. From the definition in Art. 31 it follows that if two points $A$, $B$ are moving in the same direction with velocities $u$, $v$ the relative velocity of $A$ with respect to $B$

in that direction is $u-v$ and that of $B$ with respect to $A$ is $v-u$.

34. If the two points are not moving in the same direction the magnitude and direction of the relative velocity of one with respect to the other may be easily obtained.

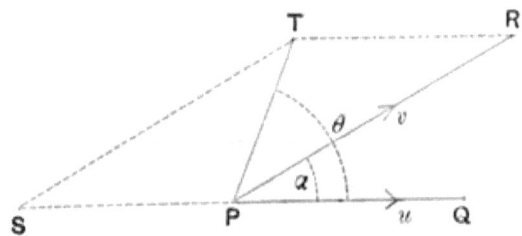

For let $PQ$, $PR$ represent the velocities $u$, $v$ of the points $A$, $B$ where $QPR = \alpha$.

Produce $QP$ to $S$ making $PS$ equal to $QP$ and complete the parallellogram $PSTR$.

Then, by Art. 31, $PT$ the resultant of the two velocities $PR$, $PS$ represents the relative velocity of $B$ with respect to $A$.

Also $PT^2 = PR^2 + RT^2 - 2PR \cdot RT \cos \alpha$
$= u^2 + v^2 - 2uv \cos \alpha$.

Also if $QPT = \theta$ we have
$$\frac{u}{v} = \frac{RT}{PR} = \frac{\sin RPT}{\sin RTP} = \frac{\sin(\theta - \alpha)}{\sin \theta},$$
so that $\tan \theta = \dfrac{v \sin \alpha}{v \cos \alpha - u}$.

Hence the relative velocity of $B$ with respect to $A$ is $\sqrt{u^2 + v^2 - 2uv \cos \alpha}$ at an angle $\tan^{-1} \dfrac{v \sin \alpha}{v \cos \alpha - u}$ with the direction of $A$.

**35.** If the components of the velocity of $A$ parallel to three fixed directions in space be $u, v, w$ and those of $B$ be $u_1, v_1, w_1$ then the components of the relative velocity of $B$ with respect to $A$ parallel to the three fixed directions are $u_1 - u$, $v_1 - v$ and $w_1 - w$.

Conversely when the velocity of a point $A$ in space is known and the velocity of another point $B$ relative to $A$ we can easily get the velocity of $B$.

For let $u, v, w$ be the component velocities of $A$ parallel to three lines fixed in space, $u_1, v_1, w_1$ the corresponding components of $B$'s velocity and $U, V, W$ those of the relative velocity of $B$ with respect to $A$.

Then we have
$$u_1 - u = U; \therefore u_1 = u + U.$$
So $\quad v_1 = v + V$, and $w_1 = w + W$.

Hence by finding the resultant of $u_1, v_1, w_1$ we have the velocity of $B$.

**36. Relative Accelerations.** If we substitute the word "acceleration" for "velocity" in Arts. 31—35, the results will still be true, since accelerations, like velocities, follow the parallelogram law.

### EXAMPLES.

1. *A train is travelling along a horizontal rail at the rate of* 60 *miles per hour, and rain is driven by the wind which is in the same direction as the motion of the train so that it falls with a velocity of* 44 *feet per second, and at an angle of* 30° *with the vertical. Find the apparent direction of the rain to a person travelling with the train.*

The true horizontal and vertical components of the rain are $44 \sin 30°$ and $44 \cos 30°$ respectively, or $22$ and $22\sqrt{3}$ feet per second.

Also the horizontal velocity of train is 60 miles per hour, or 88 feet per second.

EXAMPLES. 23

∴ the components of the velocity of the rain relative to the train are $-66$ and $22\sqrt{3}$ feet per second.

The resultant of these is a velocity of $44\sqrt{3}$ feet per second, at an angle $-\tan^{-1}\dfrac{66}{22\sqrt{3}}$ or $-60°$ with the vertical.

Hence the rain appears to meet the train at an inclination of $60°$ to the vertical and the real direction of the rain is at right angles to the apparent direction.

2. A railway train moving at the rate of 30 miles per hour is struck by **a stone** moving horizontally and **at right angles** to the train with a velocity of 33 feet per second. Find the magnitude and direction of the velocity with which the stone appears to meet the train.

*Ans.* 55 feet per second at an angle $\tan^{-1}(-\tfrac{3}{4})$ with the direction of the train's motion.

3. One ship is sailing **south** with a velocity of $15\sqrt{2}$ miles per hour, and another south east at the rate **of** 15 miles per **hour**. Find the apparent velocity and direction of motion of the second vessel to an **observer on the** first vessel.

*Ans.* **15 miles per** hour in a north **east** direction.

4. In a tunnel, drops of water which are falling from the roof are noticed to pass the carriage window in a direction making an angle $\tan^{-1}\tfrac{1}{2}$ with the horizon, and they are known to have a velocity of 24 feet per second. Neglecting **the** resistance of the air find the velocity of the train.

*Ans.* $32\tfrac{8}{11}$ miles per hour.

5. **A ship is** sailing due east, and it is known that the wind is blowing **from the** north-west, and the apparent direction of the wind (as shewn by a vane on the mast of the ship) is from N.N.E.; shew that the velocity **of** the ship is equal to that of the wind.

6. **A** person travelling eastward at the rate of **4** miles per hour, finds that the wind seems **to** blow directly from the north; on doubling his speed it appears to come from the north east; find the direction of the wind and its velocity.

*Ans.* $4\sqrt{2}$ miles per hour; towards S.E.

7. A person travelling toward the north east, finds that the wind appears to blow from the north, but when he doubles his speed it seems

to come from a direction inclined at an angle $\cot^{-1} 2$ on the east of north. Find the true direction of the wind.

*Ans.* Towards the east.

8. A railway train is moving at the rate of 28 miles per hour, when a pistol shot strikes it in a direction making an angle $\sin^{-1}\frac{3}{5}$ with the train. The shot enters one compartment at the corner furthest from the engine and passes out at the diagonally opposite corner; the compartment being 8 feet long and 6 feet wide, shew that the shot is moving at the rate of 84 miles per hour nearly, and traverses the carriage in $\frac{5}{77}$ths of a second.

9. Two points move simultaneously from points $A$ and $B$ which are 5 feet apart, one from $A$ towards $B$ with a velocity which would cause it to reach $B$ in 3 seconds, and the other at right angles to the former with $\frac{3}{4}$ of its velocity. Find their relative velocity in magnitude and direction, the shortest distance between them, and the time when they are nearest.

*Ans.* The relative velocity of the 2nd with respect to the 1st is $2\frac{1}{12}$ ft.-sec. units at an $\angle \tan^{-1}\frac{3}{4}$ with $BA$; the shortest distance is 3 feet, at the end of $1\frac{23}{25}$ secs.

10. Two points move with velocities $v$ and $2v$ respectively in opposite directions in the circumference of a circle. In what position is their relative velocity greatest and least and what value has it then?

11. The radius of the earth being 4000 miles, the latitude $\lambda$ of a point on the earth's surface at which a train travelling westward at the rate of a mile per minute is at rest in space is given by $\cos\lambda = \dfrac{9}{50\pi}$.

### 37. Angular Velocity.

Suppose we have a point $P$ in motion in a plane, and let $O$ be a fixed point in that plane and $Ox$ a fixed line through $O$, then the rate at which the angle $xOP$ increases is called the angular velocity of the point $P$ about $O$.

If the point move from $P$ to $Q$ in time $\tau$ and if $\theta$ be the number of radians in the angle $POQ$, then the angular velocity of $P$ about $O$ is the limit of $\dfrac{\theta}{\tau}$ when $\tau$, and therefore $\theta$, becomes indefinitely small.

## ANGULAR VELOCITY.

Let $V$ be the linear velocity of the moving point, so that $PQ = V \cdot \tau$ (when $\tau$ is small). Also let $OY$ be the perpendicular from $O$ on the line $PQ$.

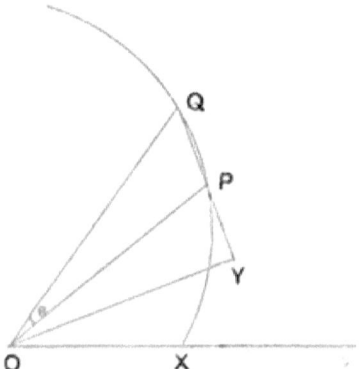

Then
$$OP \cdot OQ \sin \theta = 2 \cdot \triangle POQ = PQ \cdot OY.$$
Hence
$$\text{Lt } \frac{\theta}{\tau} = \text{Lt } \frac{\theta}{\sin \theta} \cdot \frac{\sin \theta}{\tau} = \text{Lt } \frac{\theta}{\sin \theta} \cdot \frac{PQ}{\tau} \cdot \frac{OY}{OP \cdot OQ} = V \cdot \frac{p}{OP^2},$$
ultimately, when $\tau$ is very small and $p$ is the perpendicular from $O$ on the tangent at $P$ to the path of the moving point.

Hence if $\omega$ be the angular velocity of the moving point about $O$ we have
$$r^2 \omega = Vp, \text{ where } OP = r.$$

38. The most important case arises when the point $P$ is describing a circle about $O$ as centre. In this case $p = r$ and we have
$$r\omega = V.$$

Hence, *If a moving point describe a circle, its angular velocity about the centre is equal to its linear velocity divided by the radius of the circle.*

This important result may be more easily obtained independently in the case of a point describing a circle.

Let $XP$ be an arc of a circle whose radius is $r$ and centre $O$, and in time $\tau$ let the point move from $P$ to $Q$.

Then, since arc $PQ = r \times \angle POQ$,

$$\therefore \omega = \frac{\angle POQ}{\tau} = \frac{1}{r} \times \frac{\text{arc } PQ}{\tau} = \frac{1}{r} \cdot V,$$

$$\therefore r\omega = V.$$

**39.** The rate of increase of the angular velocity is called the angular acceleration. It is measured in the same way as linear acceleration.

**40. Areal Velocity.** If the path of the moving point $P$ meet a fixed line $Ox$ in $X$ then the rate of increase of the area $XOP$ is called the areal velocity of $P$.

As before, the areal velocity

$$= \text{the limit of the quantity } \frac{\text{area } POQ}{\tau}.$$

Now area $POQ = \tfrac{1}{2} PQ \times$ perpendicular from $O$ on $PQ$.

$\therefore$ areal velocity $= \tfrac{1}{2}$ Lt $\dfrac{PQ}{\tau} \times$ perpendicular from $O$

$$= \tfrac{1}{2} Vp = \tfrac{1}{2} r^2 \omega. \quad \text{(Art. 37.)}$$

Similarly the rate of increase of the areal velocity is called the areal acceleration.

### EXAMPLES.

**1.** *A wheel rolls uniformly on the ground, without sliding, its centre describing a straight line; to find the velocity of any point of its rim.*

Let $O$ be the centre, and $r$ the radius of the wheel, and let $v$ be the velocity with which the centre advances. Let $A$ be the point of the wheel in contact with the ground at any instant.

# EXAMPLES.

Now the wheel turns uniformly round its centre whilst the centre moves forward in a straight line; also, since each point of the wheel in succession touches the ground, it follows that any point of the wheel

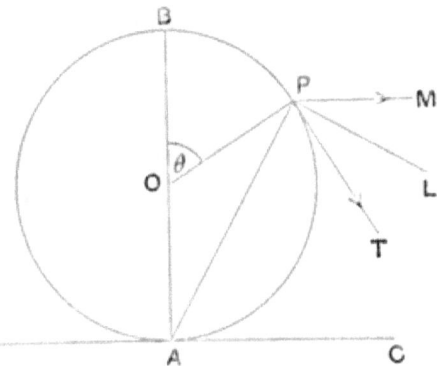

describes the perimeter of the wheel relative to the centre, whilst the centre moves through a distance equal to the perimeter; hence the velocity of any point of the wheel relative to the centre is equal in magnitude to the velocity $v$ of the centre. Hence any point $P$ of the wheel possesses two velocities each equal to $v$, one along the tangent, $PT$, at $P$ to the circle, and the other in the direction in which the centre $O$ is moving.

Hence velocity of $A = v - v = 0$, and so $A$ is at rest for the instant.

So velocity at $B = v + v = 2v$.

Consider the motion of any other point $P$. It has two velocities each equal to $v$ along $PM$ and $PT$ respectively.

Now the $\angle MPT = \angle POB = \theta$ (say).

The resultant of these two velocities $v$ is a velocity $2v \cos \dfrac{\theta}{2}$ along $PL$ where
$$\angle LPT = \frac{\theta}{2} = \angle OPA.$$

Hence $\angle APL = OPT =$ a right angle.

∴ direction of motion of the point $P$ is perpendicular to $AP$, and its angular velocity about $A$

$$= \frac{2v \cos \dfrac{\theta}{2}}{AP} = \frac{2v \cos \dfrac{\theta}{2}}{2r \cos \dfrac{\theta}{2}} = \frac{v}{r}$$

$=$ the angular velocity of the wheel about $O$.

Hence each point of the wheel is turning about the point of contact of the wheel with the ground, with a constant angular velocity whose measure is the velocity of the centre of the wheel divided by the radius of the wheel.

2. A wheel turns about its centre making 200 revolutions per minute; what is the angular velocity of any point on the wheel about the centre?

*Ans.* $\tfrac{20}{3}\pi$.

3. A point moves in a circle with uniform speed; shew that its angular velocity about any point on the circumference is constant.

4. A point describes uniformly a given straight line; shew that its angular velocity about a fixed point varies inversely as the square of its distance from the fixed point.

5. A point is describing a parabola with uniform speed; shew that its angular velocity about the focus, $S$, at any point, $P$, varies inversely as $SP^{\frac{3}{2}}$.

6. A point moves in an ellipse so that its angular velocity about one focus varies inversely as the square of the radius vector; shew that the angular velocity about the other focus varies inversely as the square of the conjugate diameter.

7. An engine is travelling at the rate of 60 miles per hour and its wheel is 4 feet in diameter; find the velocity and direction of motion of each of the two points of the wheel which are at a height of 3 feet above the ground.

*Ans.* $88\sqrt{3}$ ft. per sec. at angles $\pm 30°$ with the horizon.

8. If a railway carriage be moving at the rate of 30 miles per hour and the diameter of its wheel be 3 feet, what is the angular velocity of the wheel when there is no sliding; and what is the relative velocity of the centre and highest point of the wheel?

*Ans.* $\tfrac{88}{3}$ radians per sec.; 44 ft. per sec.

**41. Theorem.** *A point moves in a straight line starting with velocity* u, *and moving with constant acceleration* f *in its direction of motion; if* v *be its velocity at the end of time* t, *when it has moved through a distance* s, *then*

(1) $\quad v = u + ft$,

(2) $\quad s = ut + \tfrac{1}{2} ft^2$,

(3) $\quad v^2 = u^2 + 2fs$.

# UNIFORMLY ACCELERATED MOTION

(1) Since $f$ denotes the acceleration i.e. the change in the velocity per unit of time, $ft$ denotes the change in the velocity in $t$ units of time.

But since the particle possessed $u$ units of velocity initially, at end of time $t$ it must possess $u + ft$ units, i.e.

$$v = u + ft.$$

(2) Again since the velocity increases uniformly throughout this interval $t$, the velocity at any time $T$ preceding the middle of this interval is as much less than the velocity at the middle of this interval as the velocity at the time $T$ after the middle of this interval is greater than the velocity at the middle of this interval.

Hence the average velocity during the interval must be the same as that at the middle of the interval, and is therefore

$$u + f \cdot \tfrac{1}{2} t.$$

∴ if $s$ be the space described,

$$s = t \left( u + f \frac{t}{2} \right).$$

$$= ut + \tfrac{1}{2} ft^2.$$

(3) The third relation can be easily deduced from the first two by eliminating $t$ between them.

For from (1), $\quad v^2 = (u + ft)^2$

$$= u^2 + 2uft + f^2 t^2$$

$$= u^2 + 2f(ut + \tfrac{1}{2} ft^2).$$

∴ by (2) $\quad v^2 = u^2 + 2fs.$

*Cor.* If the point move in a *curved* line, starting with velocity $u$, and moving with a constant acceleration $f$ which at each instant of its motion is directed along the tangent to its path, the above relations will still be true.

**42.** *Alternative proof of equation* (2).

Let the time $t$ be divided into $n$ equal portions of time each equal to $\tau$ so that $t = n\tau$.

Then the velocities of the point at the beginning of each of these successive intervals are

$$u,\ u+f\tau,\ u+2f\tau,\ \ldots\ldots\ u+(n-1)f\tau.$$

Hence the space $s_1$ which *would be* moved through by the particle, if it moved during each of these intervals $\tau$ with the velocity which it has at the *beginning* of each, is

$$s_1 = u\cdot\tau + [u+f\tau]\cdot\tau + \ldots\ldots + [u+f(n-1)\tau]\cdot\tau$$
$$= nu\tau + f\tau^2\cdot\{1+2+3\ldots\ldots+(n-1)\}$$
$$= nu\tau + f\tau^2\cdot\frac{n(n-1)}{1\cdot2}$$
$$= ut + \tfrac{1}{2}ft^2\left(1-\frac{1}{n}\right),\ \text{since}\ \tau = \frac{t}{n}.$$

Also the velocities at the end of each of these intervals are

$$u+f\tau,\ u+2f\tau,\ \ldots\ldots\ u+nf\tau.$$

Hence the space $s_2$ which *would be* moved through by the point if it moved during each of these intervals $\tau$ with the velocity it has at the *end* of each is

$$s_2 = (u+f\tau)\cdot\tau + (u+2f\tau)\cdot\tau + \ldots\ldots + (u+nf\tau)\cdot\tau$$
$$= nu\tau + f\tau^2(1+2+3\ldots\ldots+n)$$
$$= ut + \tfrac{1}{2}ft^2\left(1+\frac{1}{n}\right),\ \text{as before.}$$

Now the true space $s$ is intermediate between $s_1$ and $s_2$; also the larger we make $n$ and therefore the smaller the intervals $\tau$ become, the more nearly do the two hypotheses approach to coincidence.

If we make $n$ infinitely large the values of $s_1$, $s_2$ both become $ut+\tfrac{1}{2}ft^2$.

Hence $\qquad\qquad\qquad s = ut + \tfrac{1}{2}ft^2.$

**43.** When the moving point starts from rest we have $u = 0$ and the formulae of Art. 41 take the simpler forms

$$v = ft,$$
$$s = \tfrac{1}{2}ft^2,$$
$$v^2 = 2fs.$$

The proofs of that article with $u$ equated to zero would apply without other change for these formulae.

## UNIFORMLY ACCELERATED MOTION.

**44.** *Space described in any particular second.*

[The student will notice carefully that the formula (2) of Art. 41 gives, not the space traversed in the $t^{th}$ second, but that traversed in $t$ seconds.]

The space described in the $t^{th}$ second
= space described in $t$ seconds − space described in $(t-1)$ seconds

$$= [ut + \tfrac{1}{2}ft^2] - [u(t-1) + \tfrac{1}{2}f(t-1)^2]$$
$$= u + \tfrac{1}{2}f[t^2 - (t-1)^2]$$
$$= u + f\frac{2t-1}{2}.$$

Hence the spaces described in the first, second, third, ..., $n$th seconds of the motion are

$$u + \tfrac{1}{2}f,\ u + \tfrac{3}{2}f,\ \ldots\ u + \frac{2n-1}{2}f.$$

These distances form an arithmetical progression whose common difference is $f$.

Hence if a body move with a uniform acceleration the distances described in successive seconds form an arithmetical progression whose common difference is equal to the number of units in the acceleration.

**45.** The space described in any particular second may be also obtained by considering the average velocity during that second.

As in Art. 41 the average velocity during the $t^{th}$ second is the same as the velocity at the middle of that second, and is therefore the velocity at the end of $(t - \tfrac{1}{2})$ seconds from the commencement of the motion.

Hence the average velocity during the $t^{th}$ second

$$= u + f(t - \tfrac{1}{2}),$$

and therefore space described in this second

$$= u + f\frac{2t-1}{2}.$$

## UNIFORMLY ACCELERATED MOTION.

### EXAMPLES.

1. In what time would a body acquire a velocity of 30 miles per hour if it started with a velocity of 4 feet per second and moved with the ft.-sec. unit of acceleration?

*Ans.* 40 secs.

2. A body starting from rest and moving with uniform acceleration describes 171 feet in the tenth second; find its acceleration.

*Ans.* 18 ft.-sec. units.

3. A point starts with a velocity of 100 cms. per second and moves with $-2$ c.g.s. units of acceleration. When will its velocity be zero, and how far will it have gone?

*Ans.* In 50 secs.; 25 metres.

4. A point moving with uniform acceleration describes in the last second of its motion $\frac{9}{25}$ ths of the whole distance. If it started from rest, how long was it in motion and through what distance did it move, if it described 6 inches in the first second?

*Ans.* 5 secs.; $12\frac{1}{2}$ ft.

5. A point moving with uniform acceleration describes 25 feet in the half second which elapses after the first second of its motion, and 198 feet in the eleventh second of its motion; find the acceleration of the point and its initial velocity.

*Ans.* Initial velocity $= 30$ ft.-sec. units; acceleration $= 16$ ft.-sec. units.

6. A point starts from rest and moves with a uniform acceleration of 18 ft.-sec. units; find the time taken by it to traverse the first, second, and third feet respectively.

*Ans.* $\frac{1}{3}$, $\frac{\sqrt{2}-1}{3}$, $\frac{\sqrt{3}-\sqrt{2}}{3}$ seconds respectively.

7. A point moves over 7 feet in the first second during which it is observed, and over 11 and 17 feet in the third and sixth seconds respectively; is this consistent with the supposition that it is subject to a uniform acceleration?

8. A point is moving in a north east direction with a velocity 6, and has accelerations 8 towards the north and 6 towards the east. Find its position after the lapse of one second. [The units are a foot and a second.]

*Ans.* Its displacement is $\sqrt{61 + 42\sqrt{2}}$ ft. at an angle $\tan^{-1} \dfrac{2+\sqrt{2}}{3}$ north of east.

# FALLING BODIES. 33

**46. Acceleration of falling bodies.** When a heavy body of any kind falls toward the earth, it is a matter of everyday experience that it goes quicker and quicker as it falls, or, in other words, that it moves with an acceleration. That it moves with a *constant* acceleration may be roughly shewn by the following experiment first performed by Morin.

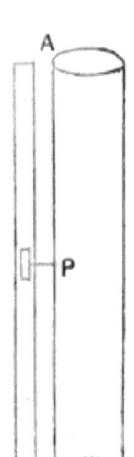

A circular cylinder covered with paper is connected with clock-work and made to rotate about its axis which is vertical. In front of the cylinder is an iron weight carrying a pencil $P$ which is compelled by guides to fall in a vertical line and is so arranged that the tip of the pencil just touches the paper on the surface of the cylinder.

When the cylinder is revolving uniformly the weight is allowed to drop and the pencil traces out a curve on the paper. When the weight has reached the ground the paper is unwrapped and stretched out on a flat surface. The curve marked out by the pencil is found to be such that the vertical distances described by the pencil from the beginning of the motion are always proportional to the squares of the horizontal distances described [i.e. the trace of the pencil is a parabola whose axis is vertical and vertex upwards]. Now since the cylinder revolved uniformly, these horizontal distances are proportional to times that have elapsed from the commencement of the motion. Hence the vertical distance described is proportional to the square of the time from the commencement of the motion.

L. D.

But from Art. 43 we know that if a point move from rest with a constant acceleration the space described is proportional to the square of the time.

Hence we infer that a falling body moves with a constant acceleration.

**47.** From the results of the foregoing and other more accurate experiments we learn that if a body be let fall towards the earth *in vacuo* it will move with an acceleration which is always the same at the same place on the earth, but which varies slightly for different places.

The value of this acceleration which is called the "acceleration due to gravity" is always denoted by the letter "$g$." The cause of this acceleration will be discussed in the next chapter.

When foot-second units are used the value of $g$ varies from about $32\cdot091$ at the equator to about $32\cdot252$ at the poles. In the latitude of London its value is about $32\cdot19$ and at any point on the earth whose latitude is $\lambda$ is about $32\cdot091\,(1 + \cdot005133\sin^2\lambda)$.

When centimetre-second units are used the extreme limits are about 978 and 983 respectively, and in the latitude of London the value is about 981.

The best method of determining the value of "$g$" is by means of pendulum experiments; we shall return to the subject again in Chapter VIII.

[*In all numerical examples **unless it is otherwise stated, the motion may be supposed to be** in vacuo **and the value of** g **taken to be** 32 **when foot-second units and** 981 **when centimetre-second units are used.*]

**48. Vertical motion under gravity.** Suppose a body is projected vertically from a point on the earth's

VERTICAL MOTION UNDER GRAVITY. 35

surface so as to start with velocity $u$. The acceleration of the body is opposite to the initial direction of motion and is therefore denoted by $-g$. Hence the velocity of the body continually gets less and less until it vanishes; the body is then for an instant at rest but immediately begins to acquire a velocity in a downward direction and retraces its steps.

**49. *Time to a given height.*** The height $h$ at which a body has arrived in time $t$ is given by substituting $-g$ for $f$ in equation (2) of Art. 41, and is therefore given by

$$h = ut - \tfrac{1}{2}gt^2.$$

This is a quadratic equation with both roots positive; the lesser root gives the time at which the body is at the given height on the way up, and the greater the time at which it is at the same height on the way down.

Thus the time that elapses before a body which starts with a velocity of 64 feet per second is at a height of 28 feet is given by

$$28 = 64t - 16t^2, \text{ whence } t = \tfrac{1}{2} \text{ or } \tfrac{7}{2}.$$

Hence the particle is at the given height in half a second from the commencement of its motion and again in 3 seconds afterwards.

**50. *Velocity at a given height and the greatest height attained.***

The velocity $v$ at a given height $h$ is by equation (3) of Art. 41 given by

$$v^2 = u^2 - 2gh.$$

Hence the velocity at a given height is independent of the time and is therefore the same at the same point whether the body is going upwards or downwards.

3—2

# VERTICAL MOTION UNDER GRAVITY.

At the highest point the velocity is just zero; hence if $x$ is the greatest height attained we have
$$0 = u^2 - 2gx.$$
$\therefore$ greatest height attained $= \dfrac{u^2}{2g}.$

Also the time $T$ to the greatest height is given by
$$0 = u - gT.$$
$$\therefore T = \dfrac{u}{g}.$$

**51.** *Velocity due to a given vertical fall from rest.*

If a body be dropped from rest its velocity after falling through a height $h$ is obtained by substituting $0$, $g$, $h$ for $u, f, s$ in equation (3) Art. 41;
$$\therefore v = \sqrt{2gh}.$$

### EXAMPLES.

1. A body is projected from the earth vertically with a velocity of 40 feet per second; find (1) how high it will go before coming to rest, (2) what times will elapse before it is at a height of 9 feet?

*Ans.* 25 ft.; ¼ sec. and 2¼ secs.

2. A body falls freely from the top of a tower and during the last second of its flight falls $\frac{16}{25}$ths of the whole distance. Find the height of the tower.

*Ans.* 100 feet.

3. A body moving in a vertical direction passes a point at a height of 54·5 centimetres with a velocity of 436 centimetres per second; with what initial velocity was it thrown up, and for how much longer will it rise?

*Ans.* 545 cms. per sec.; ⅔th sec.

4. A particle passes a given point moving downwards with a velocity of fifty metres per second; how long before this was it moving upwards at the same rate?

*Ans.* 10·2 secs.

## EXAMPLES. 37

5. A body projected vertically downwards described 720 feet in $t$ seconds, and 2240 feet in $2t$ seconds; find $t$ and the velocity of projection.

*Ans.* $t = 5$; velocity of projection = 64 ft. per sec.

6. A body is projected upwards with a certain velocity, and it is found that when in its ascent it is 960 feet from the ground it takes 4 seconds to return to the same point again; find the velocity of projection and the whole height ascended.

*Ans.* 256 ft. per sec.; 1024 ft.

7. A tower is 288 feet high; one body is dropped from the top of the tower and at the same instant another projected vertically upwards from the bottom, and they meet half-way up; find the initial velocity of the projected body and its velocity when it meets the descending body.

*Ans.* 96 ft. per sec.; zero.

8. A body is dropped from the top of a given tower, and at the same instant a body is projected, in the same vertical line, with a velocity which would be just sufficient to take it to the same height as the tower; find where they will meet.

*Ans.* The first body will have fallen from the top through one quarter the height of the tower.

9. A body is projected upwards with velocity $u$, and $t$ seconds afterwards another body is similarly projected with the same velocity; find when, and where, they will meet.

10. A stone is dropped into a well and the sound of the splash is heard in $7\frac{7}{10}$ seconds; if the velocity of sound be 1120 feet per second, find the depth of the well.

*Ans.* 784 ft.

11. A stone is dropped into a well and reaches the bottom with a velocity of 96 feet per second, and the sound of the splash on the water reaches the top of the well in $3\frac{1}{5}$ seconds from the time the stone starts; find the velocity of sound.

12. A balloon ascends with a uniform acceleration of 4 ft.-sec. units; at the end of half a minute a body is released from it; find the time that elapses before it reaches the ground.

*Ans.* 15 secs.

13. A cage in a mine shaft descends with 2 ft.-sec. units of acceleration. After it has been in motion for 10 seconds, a particle is dropped on it from the top of the shaft. What time elapses before the particle hits the cage?

*Ans.* $3\frac{1}{3}$ secs.

14. Assuming the acceleration of a falling body at the surface of the moon to be one-sixth of its value on the earth's surface, find the height to which a particle will rise if it be projected vertically upward from the surface of the moon with a velocity of 40 feet per second.

*Ans.* 150 ft.

**52. *Motion down a smooth inclined plane.***

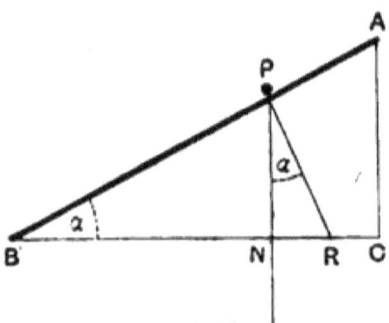

Let $AB$ be a smooth inclined plane inclined at a given angle $\alpha$ to the horizon and let $P$ be a body on it.

If there were no plane to stop its motion the body would fall vertically with an acceleration $g$.

Now, by the parallelogram of accelerations, a vertical acceleration $g$ is equivalent to

(1) an acceleration $g \cos \alpha$ perpendicular to the plane in the direction $PR$,

and (2) an acceleration $g \sin \alpha$ down the plane.

The plane prevents any motion perpendicular to itself.

Hence the body moves down the plane with an acceleration $g \sin \alpha$, and the investigation of its motion is similar to that of a freely falling body, except that instead of $g$ we have to substitute $g \sin \alpha$ in the equation of Art. 51.

It follows at once that the velocity acquired in sliding down a length $l$ of the plane is

$$\sqrt{2g \sin \alpha \cdot l} = \sqrt{2g \cdot l \sin \alpha} = \sqrt{2g \cdot PN},$$

and is therefore the same as that acquired by a particle in falling freely through a vertical height equal to that of the plane.

53. If the body be projected up the plane with initial velocity $u$ an investigation similar to that of Arts. 49 and 50 will give the motion. The greatest distance attained measured up the plane is $\dfrac{u^2}{2g \sin \alpha}$; the time taken in traversing this distance is $\dfrac{u}{g \sin \alpha}$ and so on.

### EXAMPLES.

1. A body is projected with a velocity of 80 feet per second up a smooth inclined plane, whose inclination is 30°; find the space described, and the time that elapses, before it comes to rest.

    *Ans.* 200 feet; 5 secs.

2. A particle slides without friction down an inclined plane and in the 5th second after starting passes over a distance of 2207·25 centimetres; find the inclination of the plane to the horizon.

    *Ans.* 30°.

3. $AB$ is a vertical diameter of a circle whose plane is vertical, $PQ$ a diameter inclined at an angle $\theta$ to $AB$. Find $\theta$ so that the time of sliding down $PQ$ may be twice that of sliding down $AB$.

    *Ans.* $\cos^{-1} \frac{1}{4}$.

4. A number of bodies slide from rest down smooth inclined planes which all commence at the same point and terminate in the same horizontal plane; shew that the velocities acquired are the same.

# UNIFORMLY ACCELERATED MOTION.

5. Two heavy bodies descend the height and length respectively of an inclined plane; shew that the times vary as the **spaces** described and that the velocities acquired are equal.

6. A heavy particle slides down a smooth inclined plane of given height; shew that the time of descent varies as the secant of the inclination of the plane to the vertical.

**54. Theorem.** *The time that a body takes to slide down any smooth chord of a vertical circle which is drawn from the highest point of the circle, is constant.*

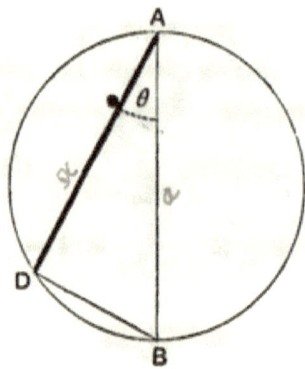

Let $AB$ be a diameter of a vertical circle of which $A$ is the highest point and $AD$ any chord.

Let $\angle DAB = \theta$; and put $AD = x$, $AB = a$, so that
$$x = a \cos \theta.$$

As in Art. 52, the acceleration down $AD$ is $g \cos \theta$. Let $T$ be the time from $A$ to $D$. Then $AD$ is the distance described in time $T$ by a particle starting from rest and moving with acceleration $g \cos \theta$.

$$\therefore x = \tfrac{1}{2} g \cos \theta \cdot T^2.$$

$$\therefore T = \sqrt{\frac{2x}{g \cos \theta}} = \sqrt{\frac{2a}{g}}.$$

LINES OF QUICKEST DESCENT. 41

This result is independent of $\theta$ and is the same as the time of falling vertically through the distance $AB$.

Hence the time of falling down any chord of this circle beginning at $A$ is the same.

55. The same theorem will be found to be true for all chords of the same circle *ending* in the *lowest* point; or if the plane of the circle be inclined to the vertical.

The theorem is also true if we substitute, for the circle, a sphere having $A$ as the highest point; for any plane section of a sphere is a circle and hence the theorem is true for all plane sections of the sphere through the vertical diameter, and hence for the sphere.

### EXAMPLES.

1. If two circles touch each other at their highest or lowest points and a straight line be drawn through this point to meet both circles, then the time of sliding from rest down the portion of this line intercepted between the two circles is constant.

2. A body slides down chords of a vertical circle ending in its lowest point; shew that the velocity on reaching the lowest point varies as the length of the chord.

56. **Lines of quickest descent.** The line of quickest descent from a given point to a curve in the same vertical plane is the straight line down which a body would slide from the given point to the given curve in the shortest *time*. It is not the same line as the geometrically shortest line that can be drawn from the given point to the curve. For example, the straight line down which the time from a given point to a given plane is least, is *not*, in general, the perpendicular from the given point upon the given plane.

## LINES OF QUICKEST DESCENT.

**57. Theorem.** *The chord of quickest descent from a given point P to a curve in the same vertical plane is PQ, where Q is a point on the curve such that a circle having P as its highest point touches the curve at Q.*

For let a circle be drawn having its highest point at $P$ to touch the given curve in $Q$. Take *any* other point $Q_1$ on the curve and let $PQ_1$ meet the circle again in $R$.

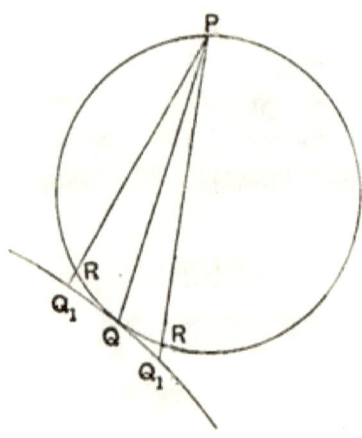

Then since $PQ_1$ is $> PR$.

∴ time down $PQ_1$ is $>$ time down $PR$.

But time down $PR =$ time down $PQ$ (Art. 54).

∴ time down $PQ_1$ is $>$ time down $PQ$,

and $Q_1$ is any point on the given curve.

∴ time down $PQ$ is less than that down any other straight line from $P$ to the given curve.

It is clear by the construction that the chord of quickest descent makes equal angles with the vertical and the normal to the curve.

*Cor.* I. If we want the chord of quickest descent from a given curve to a given point $P$, we must describe

# EXAMPLES. 43

a circle having the given point $P$ as its lowest point to touch the curve in $Q$; then $QP$ is the required straight line.

*Cor.* II. If instead of a curve we have to find the chord of quickest descent **from a point to a** given surface, we must describe **a** sphere **having the** given point **as highest point and touching the given** surface; then **the straight line from the given point to the point of contact of the sphere is the chord required.**

## EXAMPLES.

1. *To find the straight line of quickest descent to a given straight line from a given point* P *in the same vertical plane.*

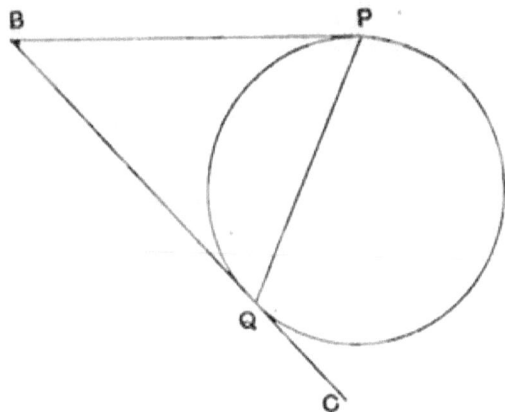

Let $BC$ be the given straight line. Then we have to describe **a circle** having its highest point at $P$ to **touch the** given straight line. **Draw** $PB$ horizontal to **meet** $BC$ **in** $B$. From $BC$ cut off a portion $BQ$ equal **to** $BP$. Then $PQ$ is the required chord; **for it is** clear that a circle **can be drawn** to touch $BP$ and $BQ$ at $P$ and $Q$ respectively.

2. *To find the line of quickest descent from a given point to a given circle in the same vertical plane.*

Join $P$ **to** the lowest point $B$ of the given circle to meet the circle again in $Q$. Then $PQ$ is the required line. For join $O$, the centre of the circle, to $Q$ and produce to meet the vertical through $P$ in $C$.

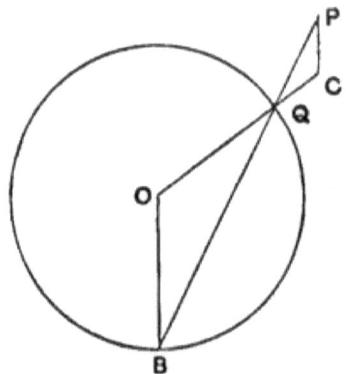

The $\angle QPC = \angle OBQ$, since $OB$ and $CP$ are parallel,
$$= \angle OQB$$
$$= \angle CQP.$$

∴ a circle whose centre is $C$, and radius $CP$, will have its highest point at $P$ and will touch the given circle at $Q$.

If $P$ be within the given circle, join $P$ to the highest point and produce to meet the circumference in $Q$; then $PQ$ will be the required line.

3. *To shew that the line of quickest descent from one curve to another in the same vertical plane makes the same angle with the normals to the curves at the points where it meets them.*

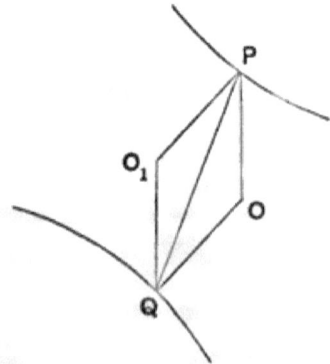

For let $PQ$ be the chord of shortest descent from one curve to the other.

EXAMPLES. 45

Let $OQ$, $O_1P$ be the normals at $Q$, $P$ and $OP$, $O_1Q$ the verticals.

Then since of all chords drawn through $P$, $PQ$ is the line of quickest descent from $P$ to the lower curve

$$\therefore \angle OQP = \angle OPQ. \quad \text{(Art. 57.)}$$

So since of all chords drawn from the upper curve to end in $Q$, $PQ$ is the line of quickest descent

$$\therefore \angle O_1QP = \angle O_1PQ.$$

But since $PO$, $QO_1$ are parallel

$$\therefore \angle OPQ = \angle O_1QP.$$

Hence  $\angle OQP = \angle O_1PQ$,

and hence the proposition is true.

It may be noticed that the line of quickest descent is the diagonal of a rhombus formed by the normals and the vertical lines drawn through its two ends.

4. *An ellipse is placed with its minor axis vertical; shew that the normal chord of quickest descent from the curve to the major axis is that drawn from a point at which the foci subtend a right angle, if there is such a point. What is the condition that there may be such a point, and what is the normal chord of quickest descent if there is not?*

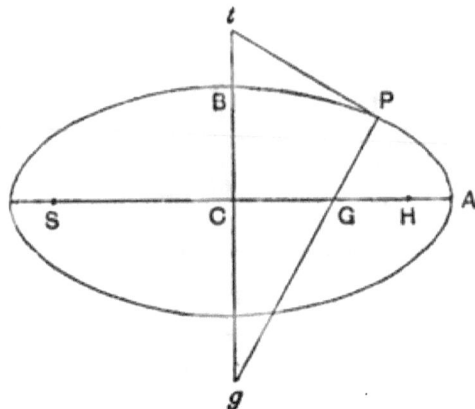

Let the tangent and normal at $P$ meet the minor axis in $t$ and $g$. Then since the ratio of $PG$ to $Pg$ is the same for all positions of $P$, the time down $Pg$ is a minimum also. But since $tPg$ is a right angle, the time down $Pg$ is the same as the time down $tg$.

∴ the time down $tg$, and hence the length $tg$, is a minimum.

But $tg$ is the diameter of a circle passing through $S$, $P$, and $H$; and the least circle passing through $S$, $H$ has its centre at $C$.

Hence we must have $SPH$ a right angle.

Again since $\angle SPH = \angle StH$, it is less than $\angle SBH$.

∴ there is no such point unless $\angle SBH$ is $> \frac{\pi}{2}$ i.e. unless $\frac{ae}{a} > \sin\frac{\pi}{4}$ or $e > \frac{1}{\sqrt{2}}$.

If $e$ be $< \frac{1}{\sqrt{2}}$ the minor axis $BC$ will be the required line.

**58. Graphic Method. Velocity-Time Curve.** *To determine the distance described in a given time when the velocity of the moving point is varying.*

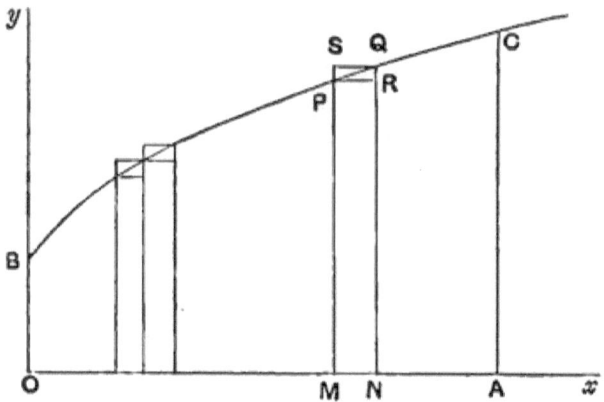

Take two straight lines $Ox$, $Oy$ at right angles to one another and let times be represented by lengths drawn along $Ox$ so that a unit of length in this direction represents a unit of time.

Let $BC$ be a line, curved or straight, such that an ordinate to it represents the velocity of the moving point at the time which is represented by the corresponding abscissa. Thus if $MP$ be an ordinate to the curve, it represents the velocity of the moving point at a time from the commencement of the motion represented by $OM$.

We shall shew that the space described by the moving point in time $OA$ is represented by the area bounded by $OB$, $OA$, $AC$ and the curved line $BC$.

Take an ordinate $QN$ close to $PM$. Then during the time $MN$ the point moves with a velocity greater than $MP$ and less than $NQ$. Hence the number of units of space it describes is intermediate between the number of units of area in the rectangles $PN$ and $QM$. Similarly if we divide $OA$ into any number of parts and erect parallelograms on each.

Hence the total space described in time $OA$ is intermediate between the space represented by the sum of the inner rectangles and that represented by the sum of the outer rectangles. Now let the number of portions into which the time $OA$ is divided be increased indefinitely in which case the area of the inner series of rectangles and that of the outer series become both ultimately equal to the area of the curve.

[For the difference between the areas of the inner and outer series of rectangles is less than the area of a rectangle whose height is $AC$ and whose base is the greatest of the parts into which $OA$ is divided, and therefore vanishes ultimately when the parts into which $OA$ is divided are taken indefinitely small. Hence the areas of the two series of rectangles are ultimately equal and therefore the area of the curve, which necessarily lies between them, must ultimately be equal to either.]

Hence the number of units of space described by the moving point is ultimately equal to the number of units of area in the area $OACB$.

*Cor.* Since $RQ$ is the increase of velocity in time $MN$, the acceleration of the moving point at this instant is equal to the limit of $\frac{RQ}{MN}$ when $MN$ is made indefinitely small.

But when $MN$ is made indefinitely small, the point $Q$ moves up to $P$, $PQ$ becomes the tangent at $P$ and $\frac{RQ}{PR}$ is ultimately the tangent of the angle that the tangent at $P$ makes with $Ox$.

Hence in the *velocity-time* curve the numerical value of the acceleration is the slope of the curve to the line along which the time is measured.

59. *Case of uniform acceleration.* Here since the acceleration is constant the slope is constant, and the velocity-time curve therefore consists of a straight line inclined at a constant angle to the time-line.

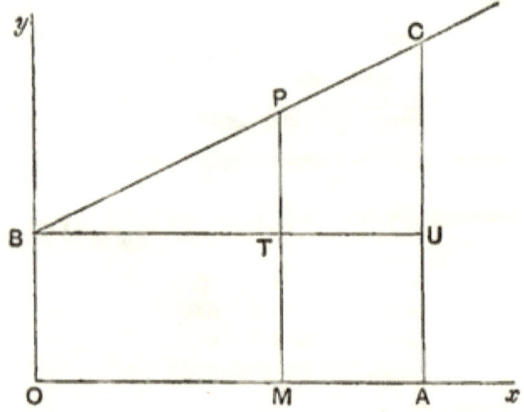

[This can be easily shown *ab initio*. For since the velocity at any time $t = u + ft$ we have
$$MP = OB + f \cdot OM.$$
$$\therefore f = \frac{MP - OB}{OM} = \tan TBP,$$
so that $TBP$ is a constant angle.]

Hence the number of **units of space** described in time $t$

= the number of units of area in $OACB$

$$= OA\left(\frac{OB + AC}{2}\right) = t\left[\frac{u + (u + ft)}{2}\right]$$

$$= ut + \tfrac{1}{2}ft^2.$$

60. **Acceleration-Time Curve.** If the ordinate to the curve drawn in Art. 58 represent the acceleration of the moving point [so that $MP$ represents the acceleration of the moving point at a time from the commencement of the motion represented by $OM$], then by a method of reasoning similar to that article, the *velocity* acquired by the moving point at the end of time $OA$ is represented by the number of units of area in $OACB$.

### EXAMPLES.

1. Draw a diagram in the case when a particle starts from rest and moves so that the product of its velocity and acceleration is constant; and find the space described in a given time.

The *velocity-time* curve in this case is such that its subnormal is constant and hence is a parabola. The space described in time $t$ is $\tfrac{2}{3}\sqrt{2\lambda}t^{\frac{3}{2}}$, where $\lambda$ is the given constant.

2. If $f \propto t$ then the velocity at any time $\propto t^2$ and the space described $\propto t^3$.

3. If $f \propto t^2$ then the velocity $\propto t^3$ and the space described $\propto t^4$.

4. If at any time the velocity of a moving point be proportional to the square of the time that has elapsed from the commencement of its motion, shew that the acceleration at any time is equal to twice the ratio of the velocity to the time that has elapsed and that the space described is one-third the product of the velocity and the time.

L. D.

## EXAMPLES.  CHAPTER I.

1. A particle is dropped from a height $h$, and after falling ⅔rds of that distance passes a particle which was projected upwards at the instant when the first was dropped. Find to what height the latter will attain.

2. A body begins to slide down a smooth inclined plane from rest at the top, and at the same instant another body is projected upwards from the foot of the plane with such a velocity that they meet half way up the plane; find the velocity of projection and determine the velocity of each when they meet.

3. Assuming that the earth in a year describes a circle uniformly about the sun as centre, that the distance between the centres is 240 radii of the sun and that the radius of the sun is 100 times that of the earth, find the velocity of the vertex of the earth's shadow taking the sun's radius as the unit of space and a year as that of time.

4. A body starts with velocity $u$ and moves with uniform acceleration; if $a$, $b$, $c$ are the spaces described in the $p$th, $q$th and $r$th seconds respectively, shew that
$$a(q-r) + b(r-p) + c(p-q) = 0.$$

5. If a space $s$ be divided into $n$ equal parts at the end of each of which the acceleration of a moving point is increased by $\dfrac{f}{n}$, find the velocity of the point after describing the distance $s$ if it started with velocity $u$.

# EXAMPLES ON CHAPTER I.

6. A particle starts from rest with acceleration $f$; at the end of time $t$ the acceleration becomes $2f$; $3f$ at end of time $2t$, and so on. Find the velocity at the end of time $nt$ and shew that the space described is
$$\frac{n(n+1)(2n+1)}{12}ft^2.$$

7. If the acceleration of a particle for the first second be $a$ and if it increase at the end of each second in geometrical progression with a common ratio $r$, shew that the space described in $n$ seconds is
$$a\left\{\frac{1}{2}\frac{r^n - 2n + 1}{r - 1} + \frac{r^n - r}{(r-1)^2}\right\}.$$

8. If a particle occupy $n$ seconds less and acquire a velocity of $m$ feet per second more at one place than at another in falling through the same distance; shew that $\dfrac{m}{n}$ equals the geometrical mean between the numerical values of gravity at the two places.

9. A train goes from rest at one station to rest at another, one mile off, being uniformly accelerated for the first $\tfrac{2}{3}$rds of the journey and uniformly retarded for the remainder, and takes 3 minutes to describe the whole distance. Find the acceleration, the retardation, and the maximum velocity.

10. An engine driver suddenly puts on his brake and shuts off steam when he is running at full speed; in the first second afterwards the train travels 87 feet, and in the next 85 feet. Find the original speed of the train, the time that elapses before it comes to rest, and the distance it will travel in this interval, assuming the brake to cause a constant retardation. Find also the time the train will take, if it is 96 yards long, to pass a spectator standing at a point 484 yards ahead of the train at the instant when the brake was applied.

11. Two points describe the same circle in such a manner that the line joining them always passes through a fixed point; shew that their velocities at any instant are proportional to their distances from the fixed point.

12. A cannon ball is moving in a direction making an angle $\theta$ with a line drawn from the ball to an observer; if $V$ be the velocity of sound and $nV$ that of the ball, shew that the whizzing of the ball will be heard in the order in which it is produced or in the reverse order according as $n$ is $\lessgtr \sec \theta$.

13. Two particles $P$, $Q$ start simultaneously from a point $A$, one sliding down a plane $AB$ inclined at an angle $a$ to the horizon, and the other falling freely; shew that their relative acceleration is $g \cos a$.

Hence shew that the line $PQ$ is always perpendicular to $AB$.

14. Find the relative motion of two points one of which revolves in a circle of radius $a$ with angular velocity $\omega$, and the other moves with velocity $a\omega$ in a straight line passing through the centre of the circle.

15. Two points describe concentric circles uniformly, the time of describing the outer circle being $m$ times that in the inner; $v$ is the speed in the former and $u$ in the latter; shew that when the angular velocity of the one relative to the other is zero the actual velocity of the one relative to the other is
$$\sqrt{\frac{m-1}{m+1}(u^2 - v^2)}.$$

16. Two points describe concentric circles of radii $a$ and $b$ with velocities varying inversely as the radii; shew that the relative velocity is parallel to the line joining the points when the angle between the radii to these points is
$$\cos^{-1} \frac{2ab}{a^2 + b^2}.$$

17. If the distance between two moving points at any time be $a$, $V$ their relative velocity and $u$, $v$ the components of $V$ in and perpendicular to the direction of $a$, shew that their distance when they are nearest to one another is $\frac{av}{V}$, and that the time before they arrive at their nearest distance is $\frac{au}{V^2}$.

18. Shew that the highest point of a wheel rolling on a horizontal plane moves twice as fast as a point on the rim whose distance from the ground is half the radius.

19. Two particles are describing in the same sense equal circles in the same plane with the same constant speed. Give a geometrical construction to find the position of the particles in which each appears to the other to be stationary, and shew that the intervals between successive stationary points are in the ratio $\pi + a - 2\theta : \pi - a + 2\theta$, where $a$ is the radius of either circle, $2c$ the distance between their centres, $a$ the angle between the central radii of the particles, and $\theta$ is given by the equation
$$c \sin\left(\frac{a}{2} - \theta\right) = a \sin \frac{a}{2}.$$

20. A body at the equator falls to the earth from relative rest. Shew that it will fall to the east of the vertical line in which the fall began, and find approximately the amount of the deviation in terms of $h$, the vertical height through which it has fallen.

21. If $O$ be the circumcentre and $O_1$ the orthocentre of a triangle $ABC$, then velocities represented by $OA$, $OB$, $OC$ are equivalent to $OO_1$; and velocities $O_1A$, $O_1B$, $O_1C$ to one represented by twice $O_1O$.

22. A particle has simultaneously impressed on it three velocities represented by $PA$, $PB$, $PC$; shew that it will move in the direction $PQ$ where $Q$ is the centroid of the triangle $ABC$.

23. $AO$, $BO$ are diameters of two circles which are in a vertical plane and touch at $O$; shew that if $POQ$ be drawn through $O$ bisecting the angle between $AOB$ and the vertical through $O$ and meeting the circles in $P$ and $Q$, then the time down $POQ$ is independent of the inclination of $AOB$ to the vertical; also if $AOB$ be vertical the time down any chord $POQ$ is constant.

24. Bodies slide down smooth faces of a pyramid starting from rest at the vertex; shew that at any time $t$ they all lie on a sphere whose radius is $\tfrac{1}{4}gt^2$.

25. A parabola is placed with its axis vertical and vertex upwards; find (1) the chord of quickest descent from the focus to the curve, (2) the focal chord of quickest descent.

26. A parabola is placed in a vertical plane with its axis inclined to the vertical. $S$ is the focus, $A$ the vertex, and $Q$ the point of the curve vertically below $S$; if $SP$ be the straight line of quickest descent from the focus to the curve, shew that
$$\angle ASP = \text{twice } \angle PSQ.$$

27. A parabola is placed with its axis horizontal; find the normal chord of quickest descent from the curve to the axis, and the chord of quickest descent from the focus to the curve.

28. An ellipse is placed with its major axis vertical; shew that the line of quickest descent from its upper focus to the curve is equal to the length of the latus rectum provided the eccentricity of the ellipse is $> \tfrac{1}{2}$.

29. An ellipse is placed with its plane vertical and minor axis horizontal; find the line of quickest descent from the curve to its centre and determine its length.

30. The time of descent from rest down chords of an ellipse passing through its lowest point is a maximum or a minimum for those chords which are parallel to the transverse and conjugate axes respectively.

31. An ellipse is placed with its major axis vertical; shew that the time down any chord of its auxiliary circle which touches the ellipse is constant.

32. A hyperbola is placed in a vertical plane with its transverse axis horizontal; shew that the time of descent down a diameter is least when the conjugate diameter is equal to the distance between the foci.

33. Shew that the time of quickest descent from a given straight line to a given circle is $\sqrt{\dfrac{2l}{g}} \sec \dfrac{\theta}{2}$, where $\theta$ is the inclination of the straight line to the horizontal, and $l$ is the shortest distance between the straight line and the circle.

34. Shew that the time of quickest descent from one circle to another in the same vertical plane is $\sqrt{\dfrac{2}{g} \dfrac{c^2 - (a+b)^2}{a+b+c\sin\alpha}}$, where $a$, $b$ are the radii, $c$ the distance between the centres, and $\alpha$ the inclination to the horizon of the line joining the centres.

35. Two parabolas are placed with axes vertical, vertices downwards, and foci coincident. Shew that there are three chords down which the time of descent of a particle under gravity is a minimum, one being the principal diameter and the other two making angles of 60° on either side of the vertical.

36. The locus of the points from which the time of descent to two given points is the same is a rectangular hyperbola.

37. Find the locus of a point from which the time of quickest descent to a given vertical circle is the same.

# CHAPTER II.

## THE LAWS OF MOTION.

61. In the last chapter we discussed a few simple cases of motion without any reference to the way in which motion is produced. In the present chapter we propose to consider the production of motion, and it will be necessary to commence with a few elementary definitions.

**Matter** is "that which can be perceived by the senses" or "that which can be acted upon by, or can exert, force."

No definition can however be given that would convey an idea of what matter is to anyone who did not already possess that idea. It, like time and space, is an elementary conception.

A **Body** is a portion of matter which is bounded by surfaces and which is limited in every direction.

A **Particle** is a portion of matter which is infinitely small in all its dimensions.

The **Mass** of a body is the quantity of matter in the body.

**Force** is that which changes, or tends to change, the state of rest or uniform motion of a body.

62. These definitions may appear to the student to be vague, but we may illustrate their meaning somewhat as follows.

If we have a small portion of any substance, say iron, resting on a smooth table we may by a push be able to move it fairly easily; if we take a larger quantity of the same iron the same effort on our part will be able to move it less easily. Again, if we take two portions of platinum and wood of exactly the same size and shape the effect produced on these two substances by equal efforts on our part will be quite different. Thus common experience shews us that the same effort applied to different bodies under seemingly the same conditions does not always produce the same result. This is because the *masses* of the bodies are different.

63. If to the same mass we apply two forces in succession, and they generate the same velocity in the same time, the forces are said to be equal.

If the same force be applied to two different masses and if it produce in them the same velocity in the same time, the masses are said to be equal.

The student will notice that we here assume that it is possible to create forces of equal intensity on different occasions, e.g. we assume that the force necessary to keep a spiral spring stretched through the same distance is always the same when other conditions are unaltered.

Hence by applying the same force in succession we can obtain a number of masses each equal to a standard unit of mass.

64. The British unit of mass is called the Imperial Pound, and consists of a lump of platinum deposited in the

Exchequer Office, of which there are in addition several accurate copies kept in other places of safety.

The French or scientific unit of mass is called a gramme, and is the one-thousandth part of a certain quantity of platinum deposited in the Archives. The gramme was meant to be defined as the mass of a cubic centimetre of pure water at a temperature of 4° C. It is a much smaller unit than a Pound.

One Gramme = about 15·432 grains.
One Pound = about 453·6 grammes.

The system of units in which a centimetre, gramme, and second, are respectively the units of length, mass and time, is generally called the C. G. S. system of units.

**65. Density.** The density of a body when uniform is the mass of a unit volume of the body; so that if $m$ be the mass of volume $V$ of a body whose density is $\rho$, then

$$m = V\rho.$$

When variable, the density at any point of a body is the ratio of the mass of a very small portion of the body surrounding the point to the volume of that portion, so that

$$\rho = \text{Limit of } \frac{m}{V}, \text{ when } V \text{ approaches to zero.}$$

66. The **Weight** of a body is the force with which the earth attracts the body.

It can be shewn that every particle of matter in nature attracts every other particle with a force which varies directly as the product of the masses of the quantities and inversely as the square of the distance between them; hence it can be deduced that a sphere attracts a particle on, or outside, its surface with a force which varies inversely as the square of the distance of the particle from the centre of the sphere. The earth

## DEFINITIONS.

is not accurately a **sphere and** therefore points on its surface are not equidistant from the centre; **hence the** attraction **of** the earth **for a given mass is** not the same at all points of its surface, and therefore the weight of a given mass is different at different points of the earth.

67. The **Momentum** or Quantity of Motion of a body is proportional to the product of the mass and the velocity of the body.

If we take as the unit of momentum the momentum of a unit mass moving with unit velocity, then the momentum of a body is $mv$, where $m$ is the mass and $v$ the velocity of the body. The direction of the momentum is the same as that of the velocity.

68. The **Vis Viva** and **Kinetic Energy** of a body are both **proportional to the** product **of** the mass and the square of the velocity of the body. If the unit of vis viva be that of a unit mass moving **with unit velocity** then the vis viva of a **body is** $mv^2$.

It is convenient **to define** the kinetic energy **of a body** as $\frac{1}{2}mv^2$, so that **it is half** its vis viva. **It will be noted that** the unit of kinetic energy is that of a mass **equal to twice** the unit of mass moving with unit **velocity.**

The kinetic **energy of a body has magnitude only, and not direction.** [Such a quantity **is called a Scalar quantity. A quantity which has magnitude and direction, such as a velocity or an** acceleration, **is called a Vector quantity.**]

69. **We can now** enunciate what are commonly called Newton's Laws of Motion. "The first two were discovered by Galileo, and the third in some of its **many forms** was known to Hooke, Huyghens, Wallis, Wren and others **before the publication of the** Principia."

## THE LAWS OF MOTION.

The Laws of Motion are;

**Law I.** *Every body continues in its state of rest, or of uniform motion in a straight line, except in so far as it be compelled by impressed force to change that state.*

**Law II.** *The rate of change of momentum is proportional to the impressed force, and takes place in the direction of the straight line in which the force acts.*

**Law III.** *To every action there is an equal and opposite reaction.*

No formal proof, experimental or otherwise, can be given of these three laws. On them however is based the whole system of Dynamics, and on Dynamics the whole theory of Astronomy. Now the results obtained, and the predictions made, from the theory of Astronomy agree so well with the actual observed facts of Astronomy that it is inconceivable that the original laws on which the subject is based should be erroneous. For example, the Nautical Almanac is published four years beforehand and the predictions in it are always correct. Hence we believe in the truth of the above three laws of motion, because the conclusions drawn from them agree with our experience.

70. **Law I.** We never see this law practically exemplified in nature, because it is impossible ever to get rid of all forces during the motion of the body. It may be seen approximately in operation in the case of a piece of dry hard ice projected along the surface of dry, well swept, ice. The only forces acting on the fragment of ice, in the direction of its motion, are the friction between the two portions of ice and the resistance of the air. The

smoother the surface of the ice the further the small portion will go, and the less the resistance of the air the further it will go. The above law asserts that **if** the ice were perfectly smooth and **if** there were no resistance of the air and no other forces acting on the body, then it would go on for ever in a straight line with uniform velocity.

The law states a principle sometimes called the *Principle of Inertia*, viz.—that a body has no *innate* tendency to change its state of rest or of uniform motion in a straight line. If a portion of metal attached to a piece of string be swung round on a smooth horizontal table, then, if the string break, the metal, having no longer any force acting on it, proceeds to move in a straight line, viz. the tangent to the circle at the point at which its circular motion ceased.

If a man step out of a rapidly moving train he is generally thrown to the ground; his feet on touching the ground are brought to rest; but, as no force acts on the upper part of his body, it continues its motion as before, and the man falls to the ground.

If a man be riding on a horse which is galloping at a fairly rapid pace and the horse suddenly stop, the rider is in danger of being thrown over the horse's head.

The above examples give good illustrations of the law and tend to elucidate its meaning.

We deduce from the law that if a body be moving uniformly in a straight line the total force acting on the body must be zero, and hence the forces, if any, which act on the body must balance one another.

71. **Law II**. From this law we derive our method of measuring force.

Let $m$ be the mass of a body, and $f$ the acceleration produced in it by the action of a force whose measure is $P$.

Then by the second law of motion

$P \propto$ rate of change of momentum,

$\propto$ rate of change of $mv$,

$\propto m \times$ rate of change of $v$ ($m$ being unaltered),

$\propto m.f.$

$\therefore P = \lambda.mf$, where $\lambda$ is some constant.

Now let the unit of force be so chosen that it may produce in unit mass the unit of acceleration.

$\therefore$ when $m = 1$ and $f = 1$, we have $P = 1$, and therefore $\lambda = 1$.

Hence, in this case, we have
$$P = m.f.$$

**72. Magnitude of this unit of force.** As explained in Art. 47 we know that when a body drops vertically *in vacuo* it moves with an acceleration which we denote by "$g$"; also the force which causes this acceleration is what we call its weight.

Hence

The weight of the unit of mass acting on the unit mass produces $g$ units of acceleration;

$\therefore \dfrac{1}{g}$. weight of unit mass acting on the unit mass produces one unit of acceleration [by the second law].

But the unit of force is that which produces unit acceleration in unit mass;

$\therefore$ the unit of force $= \dfrac{1}{g} \times$ weight of unit mass.

73. In our ordinary system of units the unit of mass is one pound and $g = 32\cdot2$ approximately. Hence the unit of force is equal to $\frac{1}{32\cdot2}$ times the weight of a pound, and is therefore equal to the weight of about half an ounce. This unit is the British Absolute Unit of Force and is called a **Poundal**, so that

$$g \,.\, \text{Poundals} = \text{weight of a lb.,}$$

where $g = 32\cdot2$ approximately.

**Hence the equation P = mf is a true relation, m being the number of pounds in the body, P the number of poundals in the force acting on it, and f the number of units of acceleration produced in the mass m by the action of the force P on it.**

This relation is sometimes expressed in the form

$$\text{Acceleration} = \frac{\text{Moving Force}}{\text{Mass moved}}.$$

N.B. All through this book the unit of force used will be a poundal unless it is otherwise stated. Thus when we say that the tension of a string is $T$ we mean $T$ poundals, so that the tension is equal to the weight of $\frac{T}{g}$ pounds.

74. **C. G. S. system of units.** In this system the unit of mass is a gramme, and $g = 981$ approximately. The unit of force in this system is therefore equal to $\frac{1}{981}$ of the weight of a gramme. This unit is the C. G. S. absolute unit of force and is called a Dyne; so that,

$$g \,.\, \text{Dynes} = \text{weight of one gramme,}$$

where $g = 981$ approximately.

Also One Poundal = about 13800 Dynes.

Hence in the C. G. S. system when we use the equation $P = mf$, the force must be expressed in dynes, and the mass in grammes.

75. A poundal and a dyne are called **Absolute Units** because their values are not dependent on the value of $g$, which varies at different places on the earth's surface. The *weight* of a pound and of a gramme do depend on this value. Hence they are called **Gravitation Units**.

76. *The weight of a body is proportional to its mass and is independent of the kind of matter of which it is composed.* The following is an experimental fact: If we have an air-tight receiver, and if we allow to drop at the same instant, from the same height, portions of matter of any kind whatever, such as a piece of metal, a feather, a piece of paper etc., all these substances will be found to have always fallen through the same distance, and to hit the base of the receiver at the same time, whatever be the substances, or the height from which they are allowed to fall.

Since these bodies always fall through the same height in the same time, therefore their velocities [rates of change of space,] and their accelerations [rates of change of velocity,] must be always the same.

Let $W_1$, $W_2$ poundals be the weights of any two of these bodies, $m_1$, $m_2$ their masses. Then since their accelerations are the same and equal to $g$, we have

$$W_1 = m_1 g,$$
$$W_2 = m_2 g;$$
$$\therefore W_1 : W_2 :: m_1 : m_2,$$

or the weight of a body is proportional to its mass.

Hence bodies whose weights are equal have equal

masses; so also the ratio of the masses of two bodies is known when the ratio of their weights is known.

N.B. The equation $W = mg$ is a numerical one and means that the number of units of force in the weight of a body is equal to the product of the number of units of mass in the mass of the body, and the number of units of acceleration produced in the body by its weight.

**77.** *Distinction between mass and weight.* The student must carefully notice the difference between the mass and the weight of a body. He has probably been so accustomed to estimate the masses of bodies by means of their weights that he has not clearly distinguished between the two. If it were possible to have a cannon ball at the centre of the earth it would have no weight there; for the attraction of the earth on a particle at its centre is zero. If however it were in motion the same force would be required to stop it as would be necessary under similar conditions at the surface of the earth. Hence we see that it might be possible for a body to have no weight; its mass however remains unaltered.

The confusion is probably to a great extent caused by the fact that the word "pound" is used in two senses which are scientifically different; it is used to denote both what we more properly call "the mass of one pound" and "the weight of one pound." It cannot be too strongly impressed on the student that, strictly speaking, a pound is a mass and a mass only; when we wish to speak of the force with which the earth attracts this mass we ought to speak of the "weight of a pound." This latter phrase is often shortened into "a pound," but care must be taken to see in which sense this word is used.

**78. Weighing by Scales and a Spring Balance.** We have pointed out (Art. 47) that the acceleration due to gravity, *i.e.* the value of $g$, varies slightly as we proceed from point to point of the earth's surface. When we weigh out a substance (say tea) by means of a pair of scales, we adjust the tea until the weight of the tea is the same as the weight of sundry pieces of metal whose masses are known, and then, by Art. 76, we know that the mass of the tea is the same as the mass of the metal. Hence a pair of scales really measures masses and not weights, and so the apparent weight of the tea is the same everywhere.

When we use a spring balance however we compare the *weight* of the tea with the *force* necessary to keep the spring stretched through a certain distance. If then we move our tea and spring balance to another place, say from London to Paris, the weight of the tea will be different whilst the force necessary to keep the spring stretched through the same distance as before will be the same. Hence the weight of the tea will pull the spring through a distance different from the former distance, and hence its apparent weight as shewn by the instrument will be different.

If we have two places, $A$ and $B$, at the first of which the numerical value of $g$ is greater than at the second, then a given mass of tea will [as tested by the spring balance] appear to weigh more at $A$ than it does at $B$.

At the centre of the earth no mass has any weight at all, so that there we could not compare the masses of bodies by means of a spring balance.

## EXAMPLES.

1. A mass of 10 pounds is placed on a smooth horizontal plane, and is acted on by a force equal to the weight of 3 pounds; find the distance described by it in 10 seconds.

Here moving force = weight of 3 lbs. = $3g$ poundals;
and mass moved = 10 pounds.

Hence, using ft.-sec. units, and supposing $g = 32$, the acceleration = $\frac{3g}{10}$.

$\therefore$ distance required = $\frac{1}{2} \cdot \frac{3g}{10} \cdot 10^2 = 480$ feet.

2. Find the magnitude of the force which, acting on a kilogramme for 5 seconds, produces in it a velocity of one metre per second.

Here velocity acquired = 100 cms. per sec.;
$\therefore$ acceleration = 20 c. g. s. units.
$\therefore$ Force = $1000 \times 20$ dynes = weight of about $\frac{1000 \times 20}{981}$ or 20·4 grammes.

3. Find the magnitude of the force which acting on a mass of 10 cwt. for 10 seconds will generate in it a velocity of 3 miles per hour.

*Ans.* Wt. of $15\frac{2}{8}$ lbs.

4. A pressure equal to the weight of a kilogramme acts on a body continuously for 10 seconds and causes it to describe 10 metres in that time; find the mass of the body.

*Ans.* 49·05 kilogrammes.

5. A horizontal pressure equal to the weight of 9 pounds acts on a mass along a smooth horizontal plane; after moving through a space of 25 feet, the mass has acquired a velocity of 10 feet per second; find its magnitude.

*Ans.* 144 lbs.

6. A body is placed on a smooth table and a pressure equal to the weight of 6 pounds acts continuously on it; at the end of three seconds the body is moving at the rate of 48 feet per second; find the mass of the body.

*Ans.* 12 lbs.

7. A body of mass 200 tons is acted on by a force equal to 112000 poundals; how long will it take to acquire a velocity of 30 miles per hour?

*Ans.* 2 min. 56 secs.

8. In what time will a force equal to the weight of 10 lbs. acting on a mass of one ton move it through 14 feet?

*Ans.* 14 secs.

## EXAMPLES.

9. A mass of 10 lbs. falls 10 feet from rest, and is then brought to rest by penetrating 1 foot into some sand; find the average pressure of the sand on it.

*Ans.* Weight of 100 lbs.

10. A certain force acting on a mass of 10 pounds for 5 seconds produces in it a velocity of 100 feet per second. Compare the force with the weight of a pound, and find the acceleration it would produce in a ton.

*Ans.* $6\frac{1}{4} : 1$ ; $\frac{5}{56}$ ft.-sec. units.

11. A train of mass 200 tons is urged forward by a force equal to the weight of $1\frac{1}{2}$ tons while the resistance it experiences is equal to the weight of one ton. How long will it take to acquire a velocity of 10 miles per hour?

*Ans.* 3 min. $3\frac{1}{3}$ secs.

12. A bullet moving at the rate of 200 feet per second is fired into a trunk of wood into which it penetrates nine inches; if a bullet moving with a similar velocity were fired into a similar piece of wood five inches thick, with what velocity would it emerge supposing the resistance to be uniform?

*Ans.* $133\frac{1}{3}$ ft.-sec. units.

13. A cannon ball of mass 1000 grammes is discharged with a velocity of 45000 centimetres per second from a cannon the length of whose barrel is 200 centimetres; shew that the mean force exerted on the ball during the explosion is $5·0625 \times 10^9$ dynes.

14. At the equator the value of $g$ is $32·09$ and in London $g = 32·2$; a merchant buys tea at the equator, at a shilling per pound, and sells in London; at what price per lb. (apparent) must he sell so that he may neither gain nor lose, if he uses the same spring balance for both transactions?

A quantity of tea which weighs one lb. at the equator will appear to weigh $\frac{32·2}{32·09}$ lbs. in London. Hence he should sell $\frac{32·2}{32·09}$ lbs. for one shilling, or at $\frac{3209}{3220}$ shillings per lb.

15. At a place $A$, $g = 32·24$ and at another place $B$, $g = 32·12$. A merchant buys goods at £10 per cwt. at $A$ and sells at $B$, using the same spring balance. What must be his charge so that he may gain 20 per cent.?

*Ans.* £12. 0s. 9d.

**79. Physical Independence of Forces.** The latter part of the Second Law states that the change of motion produced by a force is in the direction in which the force acts.

Suppose we have a particle in motion in the direction $AB$ and a force acting on it in the direction $AC$; then the law states that the velocity in the direction $AB$ is unchanged, and that the only change of velocity is in the direction $AC$; so that to find the real velocity of the particle at the end of a unit of time, we must compound the velocity in the direction $AB$ with the velocity generated in that unit of time by the force in the direction $AC$. The same reasoning would hold if we had a second force acting on the particle in some other direction, and so for any system of forces. Hence if a set of forces act on a particle at rest, or in motion, their combined effect is found by considering the effect of each force on the particle just as if the other forces did not exist and as if the particle were at rest, and then compounding these effects. This principle is often referred to as that of the *Physical Independence of Forces*.

As an illustration of this principle consider the motion of a ball allowed to fall from the hand of a passenger in a train which is travelling rapidly. It will be found to hit the floor of the carriage at exactly the same spot as it would have done if the carriage had been at rest. This shews that the ball must have continued to move forward with the same velocity that the train had or, in other words, the weight of the body only altered the motion in the vertical direction, and had no influence on the horizontal velocity of the particle.

80. **Parallelogram of Forces.** We have shewn in Art. 28 that if a particle of mass $m$ have accelerations $f_1$, $f_2$ represented in magnitude and direction by lines $AB$, $AC$, then its resultant acceleration $f_3$ is represented in magnitude and direction by $AD$, the diagonal of the parallelogram of which $AB$, $AC$ are adjacent sides.

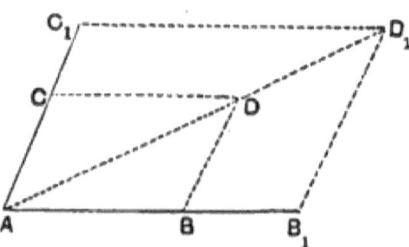

Since the particle has an acceleration $f_1$ in the direction $AB$ there must be a force $P_1 (= mf_1)$ in that direction, and similarly a force $P_2 (= mf_2)$ in the direction $AC$. Let $AB_1$, $AC_1$ represent these forces in magnitude and direction. Complete the parallelogram $AB_1D_1C_1$. Then since the forces in the directions $AB_1$, $AC_1$ are proportional to the accelerations in those directions,

$$\therefore AB_1 : AB :: B_1D_1 : BD.$$

Hence by simple geometry we have $A$, $D$, $D_1$ in a straight line, and

$$\therefore AD_1 : AD :: AB_1 : AB.$$

$\therefore AD_1$ represents the force which produces the acceleration represented by $AD$, and hence is the force which is equivalent to the forces represented by $AB_1$, $AC_1$.

Hence we infer the truth of the Parallelogram of Forces which may be enunciated as follows;

*If a particle be acted on by two forces represented in magnitude and direction by the two sides of a parallelogram drawn from a point, they are equivalent to a force represented in magnitude and direction by the diagonal of the parallelogram passing through the point.*

*Cor.* If in Arts. 17—23 which are founded on the Parallelogram of Velocities we substitute the word "force" for "velocity" they will still be true.

81. The student will notice that since a force has direction as well as magnitude it, like a velocity or an acceleration, is a vector quantity. All vector quantities will be found to be compounded according to the parallelogram law, whilst scalar quantities are compounded by simple addition.

82. **Law III.** *To every action there is an equal and opposite reaction.*

Every exertion of force consists of a mutual action between two bodies. This mutual action is called the stress between the two bodies, so that the Action and Reaction of Newton together form the Stress.

If a book rest on a table, the book presses the table with a force equal and opposite to that which the table exerts on the book.

If a man raise a weight by means of a string tied to it, the string exerts on the man's hand exactly the same force that it exerts on the weight, but in the opposite direction.

The attraction of the earth on a body is its weight, and the body attracts the earth with a force equal and opposite to its own weight.

When a horse drags a canal-boat by means of a rope, the rope drags the horse back with a force equal to that with which it drags the boat forward.

## 83. Motion of two particles connected by a string passing over a small smooth pulley.

*Two particles of masses* $m_1$, $m_2$ *are connected by a light inextensible string which passes over a small smooth pulley. If* $m_1$ *be* $> m_2$, *find the resulting motion of the system, and the tension of the string.*

Let the tension of the string be $T$ poundals; the pulley being smooth, this will be the same throughout the string.

Since the string is inextensible, the velocity of $m_2$ upwards must, throughout the motion, be the same as that of $m_1$ downwards.

Hence their accelerations [rates of change of velocity] must be the same in magnitude. Let the magnitude of the common acceleration be $f$.

Now the force on $m_1$ downwards is $m_1 g - T$ poundals. Hence
$$m_1 g - T = m_1 f \quad \text{......(i).}$$
So the force on $m_2$ upwards is $T - m_2 g$ poundals;
$$\therefore T - m_2 g = m_2 f \quad \text{......(ii).}$$

Adding (i) and (ii) we have $f = \dfrac{m_1 - m_2}{m_1 + m_2} g$, giving the common acceleration.

Also from (ii) $T = m_2 (f + g) = \dfrac{2 m_1 m_2}{m_1 + m_2} g$ poundals.

Since the acceleration is known and constant, the equations of Art. 41 give the space moved through and the velocity acquired in any given time.

*Cor.* 1. If the tension of the string is equal to the weight of $M$ lbs., or to $Mg$ poundals, we have
$$M = 2 \frac{m_1 m_2}{m_1 + m_2}.$$

Hence $\dfrac{2}{M} = \dfrac{1}{m_1} + \dfrac{1}{m_2}$, and the tension of the string is therefore equal to the weight of a mass, which is the harmonic mean between the two masses.

*Cor.* 2. Since the harmonic mean between two quantities is always less than the arithmetic mean, it follows that the tension of the string is less than the semi-sum of the weights of the masses. Hence the pressure on the axle of the pulley, which is equal to twice the tension of the string, is less than the sum of the weights of the masses.

84. **Atwood's machine.** This machine in its simplest form consists of a vertical pillar $AB$, of about 8 feet in height, firmly clamped to the ground, and carrying at its top a light pulley which will move very freely. This pillar is graduated and carries two platforms $D$, $F$, and a ring $E$, all of which can be affixed by screws at any height desired. The platform $D$ can also be instantaneously dropped. Over the pulley passes a fine cord supporting at its ends two long thin equal weights, one of which, $P$, can freely pass through the ring $E$. Another small weight $Q$ is provided, which can be laid upon the weight $P$, but which cannot pass through the ring $E$.

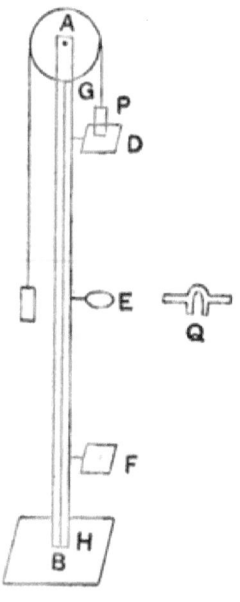

The weight $Q$ is laid upon $P$, the platform $D$ is dropped, and motion ensues; the weight $Q$ is left behind as the weight $P$ passes through the ring; the weight $P$ then traverses the distance $EF$ with constant velocity, and the time $T$ which it takes to describe this distance is carefully measured.

By the last article, the acceleration of the system as the weight falls from $D$ to $E$ is $\dfrac{(Q+P)-P}{(Q+P)+P}g$ or $\dfrac{Q}{Q+2P}g$. Denote this by $f$ and let $DE = h$.

Then the velocity $v$ on arriving at $E$ is given by
$$v^2 = 2fh.$$

After passing $E$, the distance $EF$ is described with constant velocity $v$.

Hence, if $EF = h_1$, we have
$$T = \frac{h_1}{v} = \frac{h_1}{\sqrt{2fh}}.$$
$$\therefore h_1^2 = \frac{2Q}{Q+2P} ghT^2.$$

Since all the quantities involved are known, this relation gives us the value of $g$.

By giving different values to $P$, $Q$, $h$, $h_1$, we can in this manner verify all the fundamental laws of motion.

In practice, the value of $g$ cannot by this method be found to any great degree of accuracy; the chief causes of discrepancy being the mass of the pulley, which cannot be neglected, the friction of the pivot on which the wheel turns, and the resistance of the air.

The effect of the pulley may be allowed for by taking the acceleration equal to $\dfrac{Q}{2P+Q+N}g$, where $N$ is a quantity to be determined by comparing the results of successive experiments.

The friction of the pivot may be minimised if its ends do not rest on fixed supports but on the circumferences of four light wheels, called friction wheels, two on each side, which turn very freely.

There are other pieces of apparatus for securing the accuracy of the experiment as far as possible, e.g. for instantaneously withdrawing the platform $D$ at the required moment, and a clock for beating seconds accurately.

### EXAMPLES.

**1.** *A particle slides down a rough inclined plane inclined to the horizon at an angle $a$; if $\mu$ be the coefficient of friction, determine the motion.*

Let $m$ be the mass of the particle so that its weight is $mg$ poundals; $R$ the normal reaction, and $\mu R$ the force of friction.

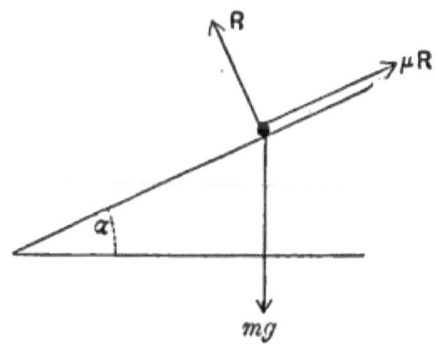

Then the total force perpendicular to the plane is
$(R - mg \cos a)$ poundals.

Also the total force down the plane is $(mg \sin a - \mu R)$ poundals.

Now perpendicular to the plane there cannot be any motion, and hence there is no change of motion.

Hence the acceleration, and therefore the force, in that direction is zero.

$$\therefore R - mg \cos a = 0 \quad\ldots\ldots\ldots\ldots\ldots\ldots\ldots\ldots\text{(i)}.$$

Also the acceleration down the plane

$$= \frac{\text{moving force}}{\text{mass moved}} = \frac{mg \sin a - \mu R}{m} = g(\sin a - \mu \cos a), \text{ by (i)}.$$

Hence the velocity of the particle after it has moved from rest over a length $l$ of the plane is by Art. 41 equal to

$$\sqrt{2gl(\sin a - \mu \cos a)}.$$

Similarly, if the particle were projected up the plane, we have to change the sign of $\mu$, and its acceleration in a direction opposite to that of its motion is $\quad g(\sin a + \mu \cos a)$.

2. *Two equally rough inclined planes, of equal height, whose inclinations to the horizon are $a_1$, $a_2$, are placed back to back; two masses $m_1$, $m_2$, are placed on their inclined faces and connected by a light inextensible string passing over a smooth pulley at the common vertex of the two planes; if $m_1$ descend, find the resulting motion.*

Let $T$ poundals be the tension of the string and $\mu$ the coefficient of friction.

Then, as in the last example, the force on $m_1$ along the plane is $m_1 g(\sin a_1 - \mu \cos a_1) - T$; and that on $m_2$ is $T - m_2 g(\sin a_2 + \mu \cos a_2)$.

Hence if $f$ be the magnitude of the common acceleration, we have

$$m_1 g(\sin a_1 - \mu \cos a_1) - T = m_1 f \ldots\ldots\ldots\ldots\ldots\ldots(i),$$
$$T - m_2 g(\sin a_2 + \mu \cos a_2) = m_2 f \ldots\ldots\ldots\ldots\ldots(ii),$$

$\therefore$ by addition

$$f = g[m_1 \sin a_1 - m_2 \sin a_2 - \mu(m_1 \cos a_1 + m_2 \cos a_2)]/(m_1 + m_2).$$

By substituting this value of $f$ in (i) we have the value of $T$.

3. *A train of mass 150 tons, moving at the rate of 30 miles per hour, comes to the foot of an incline of 1 in 100, and the engine exerts a constant force on the train until it arrives at the top of the incline, which is one mile in length, when the steam is shut off, and the train runs half a mile before coming to rest. Assuming the total resistance due to friction and the resistance of the air to be 8 lbs. per ton, find the force exerted by the engine.*

When we say that the resistance is 8 lbs. per ton we mean that for every ton in the mass of the train the resistance is equal to the *weight* of 8 pounds.

$\therefore$ total resistance = weight of 1200 lbs. = $1200g$ poundals.

The resolved part of the weight down the plane

$$= \frac{150 \times 2240 \times g}{100} \text{ poundals} = 3360g \text{ poundals}.$$

EXAMPLES. 77

Let the force exerted by the engine be equal to the weight of $x$ tons or to $2240 \times x \times g$ poundals.

∴ total acceleration up the plane

$$= (2240x - 3360 - 1200)g \div 150 \cdot 2240$$

$$= g\left(\frac{x}{150} - \frac{19}{1400}\right) = f \text{ (say)}.$$

Also the velocity at the foot of the incline is 44 feet per second.

Let $v$ be the velocity on arriving at the top of the incline. Then by Art. 41

$$v^2 = 44^2 + 2f \cdot 1760 \cdot 3 \ldots\ldots\ldots\ldots\ldots\ldots\ldots\text{(i)}.$$

On the level at the top the acceleration is $-\dfrac{1200g}{150 \cdot 2240}$ or $-\dfrac{g}{280}$. Also after running half a mile the velocity is zero.

$$\therefore 0 = v^2 - 2\frac{g}{280} \cdot 880 \cdot 3 \ldots\ldots\ldots\ldots\ldots\ldots\text{(ii)}.$$

Equating the two values of $v$ in (i), (ii) we get $x = 1\frac{199}{418}$.

Hence the tractive force is equal to the weight of $1\frac{199}{418}$ tons.

4. *A body whose mass is* n *pounds is placed on a horizontal plane which is in motion in a vertical direction; if the pressure of* **the** *body on the plane be equal to the weight of* m *pounds, find* **the** *acceleration with which the plane is moving.*

Since the pressure of the body **on** the plane is equal to the weight of $m$ **lbs.** or to $mg$ poundals, therefore by the third law of motion the pressure of the plane on the body is $mg$ poundals in the opposite direction.

The only other force acting on the body **is its** weight, which is $ng$ poundals downwards.

∴ the resultant force **on the body is** $(m-n)g$ **poundals vertically** upwards.

Hence **the** acceleration **of** the body

$$= \frac{\text{moving force}}{\text{mass moved}} = \frac{m-n}{n}g \text{ vertically upwards}.$$

**Cor.** If $m$ be $>n$, i.e. if the pressure of the body on the plane be greater than its weight, **this acceleration** is positive; hence the acceleration of the plane is vertically upwards, and **it is** either moving upwards with an increasing velocity or downwards with a decreasing velocity.

If $m$ be $<n$, this **acceleration is negative;** hence the acceleration of the plane is downwards, and it is either moving upwards with a decreasing velocity or downwards with an increasing velocity.

# EXAMPLES.

5. A mass of 9 lbs. descending vertically drags up a mass of 6 lbs. by means of a string passing over a smooth pulley; find the acceleration of the system and the tension of the string.

*Ans.* $\frac{g}{5}$; weight of $7\frac{1}{5}$ lbs.

6. Two masses each equal to $m$ are connected by a string passing over a smooth pulley; what mass must be taken from one and added to the other so that the system may describe 200 feet in 5 seconds?

*Ans.* $\frac{m}{2}$.

7. Two equal masses, of 3 lbs. each, are connected by a light string hanging over a smooth peg; if a third mass of 3 lbs. be laid on one of them, by how much is the pressure on the peg increased?

*Ans.* By the weight of 2 lbs.

8. A mass of 4 ozs. is attached by a string passing over a smooth pulley to a larger mass; find the magnitude of the latter so that, if after the motion has continued 3 seconds the string be cut, the former will ascend $\frac{1}{9}$ ft. before descending.

*Ans.* 5 ozs.

9. Two scalepans of mass 3 lbs. each are connected by a string passing over a smooth pulley; shew how to divide a mass of 12 lbs. between the two scalepans so that the heavier may descend a distance of 50 feet in the first five seconds.

*Ans.* The mass must be divided in the ratio 19 : 13.

10. Two strings pass over a smooth pulley; on one side they are attached to masses of 3 and 4 lbs. respectively, and on the other to one of 5 lbs.; find the tensions of the strings and the acceleration of the system.

*Ans.* Weights of $2\frac{1}{2}$ and $3\frac{1}{3}$ lbs. respectively; $\frac{g}{6}$.

11. A string hung over a pulley has at one end a mass of 10 lbs. and at the other end masses of 8 and 4 lbs. respectively; after being in motion for 5 seconds the four pound mass is taken off; find how much further the masses go before they first come to rest.

*Ans.* 29 ft. 9 ins. nearly.

12. A mass $m$ pulls a mass $m'$ up an inclined plane, inclination $\alpha$, by means of a string passing over a pulley at the top of the plane; shew that the acceleration is

$$\frac{m - m' \sin \alpha}{m + m'} g.$$

EXAMPLES. 79

13. A mass of 12 lbs. drags a mass of 16 lbs. up an inclined plane, whose inclination is 30°, being attached to it by means of a string passing over the top of the plane; find the distance described in 5 seconds and the tension of the string.

*Ans.* 57½ feet; weight of 10⅔ lbs.

14. A mass of 6 ounces slides down a smooth inclined plane, whose height is half its length, and draws another mass from rest over a distance of three feet in five seconds along a horizontal table which is level with the top of the plane over which the string passes; find the mass on the table.

*Ans.* 24 lbs. 10 ozs.

15. A mass of 5 lbs. on a rough horizontal table is connected by a string with a mass of 8 lbs. which hangs over the edge of the table; find the coefficient of friction in order that the heavier mass may move vertically with half the acceleration it would have if it fell freely.

*Ans.* $\mu = \cdot 3$.

16. A mass $Q$ on a rough horizontal plane, coefficient of friction $\sqrt{3}$, is connected by a string with a mass $3Q$ which hangs over the edge of the table; if motion ensue and the string break after four seconds, find how far the new position of equilibrium of $Q$ will be from its initial position.

*Ans.* 96 feet.

17. Sixteen balls of equal mass are strung like beads on a string; some are placed on an inclined plane of inclination $\sin^{-1} \frac{1}{4}$, and the rest hang over the top of the plane; how have the balls been arranged if the acceleration at first is $\dfrac{g}{2}$?

*Ans.* 10 balls must hang vertically.

18. A mass $P$ is drawn up a smooth plane inclined at an angle of 30° to the horizon by means of a mass $Q$ which descends vertically, the masses being joined by a string passing over a smooth pulley at the top of the plane; if the acceleration of the system be one-fourth that of a freely falling body find the ratio of $Q$ to $P$.

*Ans.* They are equal.

19. A man whose weight is 8 stone stands on a lift which moves with a uniform acceleration of 12 ft.-sec. units; find the pressure on the floor when the lift is (1) ascending, (2) descending.

*Ans.* (1) 154 lbs. weight; (2) 70 lbs. weight.

# 80 EXAMPLES.

20. A balloon ascends with a uniformly accelerated velocity, so that a mass of one cwt. produces on the floor of the balloon the same pressure which 116 lbs. would produce on the earth's surface; find the height which the balloon will have attained in one minute from the time of starting.

*Ans.* $2057\frac{1}{2}$ ft.

21. A bucket containing one cwt. of coal is drawn up the shaft of a coal-pit and the pressure of the coal on the bottom of the bucket is equal to the weight of 126 lbs. Find the acceleration of the bucket.

*Ans.* $\frac{g}{8}$ ft.-sec. units.

22. A bullet with an initial velocity of 1500 feet per second strikes a target at 1200 yards distance with a velocity of 900 feet per second; the range of the bullet being supposed horizontal, compare the mean resistance of the air with the weight of the bullet.

*Ans.* They are as $25 : 4$.

23. A train of mass 200 tons is running at the rate of 40 miles per hour down an incline of 1 in 120; find the resistance necessary to stop it in half a mile.

*Ans.* The weight of $5\frac{3}{11}$ tons.

24. A train runs from rest for one mile down a plane whose descent is one foot vertically for each 100 feet of its length; if the resistances be equal to 8 lbs. per ton how far will the train be carried along the horizontal level at the foot of the incline?

*Ans.* 1 mile 1408 yards.

25. $P$ hangs vertically and is 9 lbs.; $Q$ is a mass of 6 lbs. on a plane whose inclination to the horizon is $30°$; shew that $P$ will drag $Q$ up the whole length of the plane in half the time that $Q$ hanging vertically would take to draw $P$ up the plane.

26. A mass $m$ is drawn up an inclined plane of height $h$ and length $l$ by means of a string passing over the vertex of the plane, from the other end of which hangs a mass $m'$. Shew that in order that $m$ may just reach the top of the plane, $m'$ must be detached after $m$ has moved through a distance

$$\frac{m+m'}{m'} \frac{hl}{h+l}.$$

27. Two masses are connected by a string passing over a small pulley; shew that if the sum of the masses be constant, the tension of the string is greater, the less the acceleration.

28. A mass $A$ draws a mass $B$ up an inclined plane by means of a string passing over the vertex; find the inclination of the plane so that $A$ may draw $B$ up a given vertical height in the least time.

*Ans.* The inclination must be $\sin^{-1}\dfrac{A}{2B}$.

29. A plane of height $h$ and inclination $a$ to the horizon has cut in it a groove inclined at an angle $\beta$ to the line of greatest slope; find the time a particle would take to describe the groove starting from rest at the top.

*Ans.* $\sqrt{\dfrac{2h}{g}} \operatorname{cosec} a . \sec \beta$.

30. A train of mass 50 tons is moving on a level at the rate of 30 miles per hour when the steam is shut off, and the brake being applied to the brake-van the train is stopped in a quarter of a mile. Find the mass of the brake-van, taking the coefficient of friction between the wheels and rails to be one-sixth and supposing the unlocked wheels to roll without sliding.

*Ans.* $6\tfrac{1}{8}$ tons.

85. **Impulse. Def.** *The impulse of a force in a given time is proportional to the product of the force (if constant, and the mean value of the force if variable) and the time during which it acts.*

If the unit of impulse be the impulse of a unit force acting for the unit of time, then the impulse of a force $F$ acting for a time $t$ is $F.t$.

The impulse of a force is also equal to the momentum generated by the force in the given time. For suppose a particle, of mass $m$, moving initially with velocity $u$ is acted on by a constant force $F$ for time $t$. If $f$ be the resulting acceleration we have $F = mf$.

But if $v$ be the velocity of the particle at the end of time $t$, we have
$$v = u + ft.$$
Hence the impulse $= Ft = mft = mv - mu$
$\phantom{.}\quad =$ the momentum generated in the given time.

The same result is true if the force be variable. For suppose the time $t$ to be divided into a very large number $(n)$ of portions each equal to $\tau$, and let these portions of time be taken so small that in each of them the force may be considered to remain appreciably constant.

Let the values of the force be $F_1, F_2, \ldots\ldots F_n$, and the resulting accelerations be $f_1, f_2 \ldots\ldots f_n$.

Then we have
$$F_1 = mf_1, \quad F_2 = mf_2, \quad \ldots\ldots\ldots \quad F_n = mf_n.$$
Hence the impulse = mean force $\times\, t$
$$= \frac{F_1 + F_2 \ldots\ldots + F_n}{n} \cdot t$$
$$= m \cdot \frac{f_1 + f_2 + \ldots\ldots + f_n}{n} \times n\tau$$
$$= m\,(f_1\tau + f_2\tau + \ldots\ldots + f_n\tau)$$
$= m \times$ sum of changes of the velocity during the whole interval
$=$ momentum generated in time $t$.

From the preceding article it follows that the second law of motion might have been enunciated in the following form;

**The change of momentum of a particle in a given time is equal to the impulse of the force which produces it and is in the same direction.**

This is the form of the second law usually quoted.

86. **Impulsive Forces.** Suppose we have a force $F$ acting for a time $\tau$ on a body whose mass is $m$, and let the velocities of the mass at the beginning and end of this time be $u$ and $v$. Then by the last article

$$F\tau = m(v - u).$$

Let now the force become bigger and bigger and the time $\tau$ smaller and smaller. Then ultimately $F$ will be almost infinitely big and $\tau$ almost infinitely small, and yet their product may be finite. For example $F$ may be equal to $10^7$ poundals, $\tau$ equal to $\frac{1}{10^7}$ seconds, and $m$ equal to one pound, in which case the change of velocity produced is the unit of velocity.

To find the whole effect of a finite force acting for a finite time we have to find two things, (1) the change in the velocity of the particle produced by the force during the time it acts, and (2) the change in the position of the particle during this time. Now in the case of an infinitely large force acting for an infinitely short time, the body moves only a very short distance whilst the force is acting on it, so that the change of position of the particle may be neglected. Hence the total effect of such a force is known when we know the change of momentum which it produces.

Such a force is called an impulsive force. Hence

**Def.** *An impulsive force is a very great force acting for a very short time, so that the change in the position of the particle during the time the force acts on it may be neglected. Its whole effect is measured by its impulse, or the change of momentum produced.*

In actual practice we never have any experience of an

infinitely great force acting for an infinitely short time. Approximate examples are, however, the blow of a hammer, the collision of two billiard balls.

The above will be true even if the force is not uniform. In this case we must define $F$ as the mean force. In the ordinary case of the collision of two billiard balls the force generally varies very considerably.

We may notice that if we have both finite and impulsive forces acting on a body at the same time the former may be neglected in comparison with the latter; for the effect of a finite force during an infinitely short time is infinitely small; a finite force necessarily requires a finite time to produce a finite effect. Hence if we are considering the effect of a collision between two balls which are in motion in the air, we can neglect the effect of gravity during the time that the collision lasts.

Ex. 1. A body, whose mass is 9 lbs., is acted on by a force which changes its velocity from 20 miles per hour to 30 miles per hour. Find the impulse of the force.

*Ans.* 132 units of impulse.

Ex. 2. A mass of 2 lbs. at rest is struck and starts off with a velocity of 10 feet per second; if the time during which the blow lasts be $\frac{1}{100}$th of a second find the average value of the force acting on the mass.

*Ans.* 2000 poundals.

Ex. 3. A glass marble whose mass is one ounce falls from a height of 25 feet and rebounds to a height of 16 feet; find the impulse and the average force between the marble and the floor if the time during which they are in contact is $\frac{1}{10}$th of a second.

*Ans.* $4\frac{1}{2}$ units of impulse; 45 poundals.

**87. Impact of two bodies.** When two masses $A$, $B$ impinge, then by the third law of motion the action of

$A$ on $B$ is, at each instant during which they are in contact, equal and opposite to that of $B$ on $A$.

Hence the impulse of the action of $A$ on $B$ is equal and opposite to the impulse of the action of $B$ on $A$.

It follows that the change in the momentum of $B$ is equal and opposite to the change in the momentum of $A$, and therefore the sum of these changes measured in the same direction is zero.

Hence the sum of the momenta of the two masses measured in the same direction is unaltered by their impact.

88. **Motion of a shot and gun.** When a gun is fired the powder is almost instantaneously converted into a gas at a very high pressure which by its expansion forces the shot out. The action of the gas is similar to that of a compressed spring trying to recover its natural position. The force exerted on the shot forwards is at any instant before the shot leaves the gun equal and opposite to that exerted on the gun backwards, and therefore the impulse of the force on the shot is equal and opposite to the impulse of the force on the gun. Hence the momentum generated in the shot is equal and opposite to that generated in the gun, if the latter be free to move.

Ex. 1. A body, of mass 3 lbs., moving with velocity 13 feet per second overtakes a body, of mass 2 lbs., moving with velocity 3 feet per second in the same straight line and they coalesce and form one body; find the velocity of this single body.

Let $V$ be the required velocity. Then since the sum of the momenta of the two bodies is unaltered by the impact, we have

$$(3+2) V = 3 \times 13 + 2 \times 3 = 45 \text{ units of momentum},$$
$$\therefore V = 9 \text{ ft. per sec.}$$

Ex. 2. A body of mass 7 lbs., moving with a velocity of 10 feet per second overtakes a body, of mass 20 lbs., moving with a velocity of 2 feet per second in the same direction as the first; if after the impact they move forward with a common velocity, find its magnitude.

*Ans.* $4\frac{2}{27}$ ft. per sec.

Ex. 3. A body, of mass 10 lbs., moving with velocity 4 feet per second meets a body of mass 12 lbs. moving in the opposite direction with a velocity of 7 feet per second; if they coalesce into one body, shew that it will have a velocity of 2 feet per second in the direction in which the larger body was originally moving.

Ex. 4. A shot of mass 1 ounce is projected with a velocity of 1000 feet per second from a gun of mass 10 lbs.; find the velocity with which the latter begins to recoil.

*Ans.* $6\frac{1}{4}$ ft. per sec.

**89. Work. Def.** *A force is said to do work when it moves its point of application.*

The measure of this work is the product of the force and the distance through which the point of application is moved in the direction of the force.

It will be noticed that this definition says nothing about the *time* occupied in changing the position of the point of application, or of the *path* by which the point of application is brought from the first into the second position.

Let a force $F$ acting in the direction $AB$ have its point of application moved from $A$ to $C$; draw $CD$ perpendicular to $AB$; then the work done by the force is proportional to the product of $F$ and $AD$. When the point $D$ is on the side of $A$ toward which the force acts then the displacement $AD$, and the work done by the force are said to be positive.

When however the point $D$ is on the opposite side as

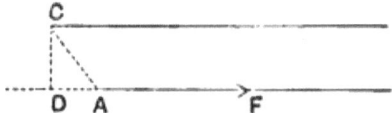

in the second figure the work done, and the displacement, are said to be negative. This is sometimes expressed by saying that the force has work done against it. If a man raise a body from the ground, the work done by the man is positive whilst that done by the body is negative.

If the angle $CAD$ be a right angle, then the point $D$ coincides with $A$, and the displacement $AD$, and therefore the work done by the force, are both zero. Hence the work done by a force when its point of application is moved in a direction perpendicular to the direction of the force is zero.

No work is done by the weight of a body which is moved about on a horizontal table.

If a body be sliding down an inclined plane no work is done by or against the pressure of the plane on the body.

90. The British absolute unit of work is the work done by a poundal in moving its point of application through one foot, and is called a **Foot-Poundal.** It is roughly equal to the work done in lifting half an ounce through one foot. With this unit the work done by a force of $F$ poundals in moving its point of application through $s$ feet is $Fs$ units of work.

The C. G. S. unit of work is that done by a dyne in moving its point of application through a centimetre, and is called an **Erg.** On account of the smallness of this unit a larger unit is sometimes used and is called a Joule. One Joule is equal to $10^7$ ergs.

One Foot-Poundal = 421390 Ergs nearly.

The practical unit used in England by engineers is called a **Foot-Pound** and is the work done in raising the weight of a pound vertically through one foot. The weight of a pound differs at different points of the earth's surface, so that this is not a suitable unit for theoretical calculations.

One Foot-Pound = $g$ . Foot-Poundals.

The corresponding unit used by French engineers is the gramme-centimetre, which is the work done in raising the weight of a gramme through one centimetre vertically.

**91. Def.** *The rate of work or* **Power** *of an agent is the amount of work that would be done by the agent in a unit of time.*

An agent is working with the British absolute unit of power when it does a foot-poundal per second.

Similarly it is working with the C. G. S. absolute unit of power when it does an erg per second. When it is performing 1 Joule or $10^7$ ergs per second it is said to be working with a power of one Watt.

The unit of power used by engineers is a **Horse-Power**. An agent is said to be working with one Horse-Power or H.-P. when it performs 33000 foot-pounds in a minute, i.e. when it raises 33000 pounds vertically through one foot per minute. This estimate of the power of a horse was made by Watt but is rather above the capacity of ordinary horses.

The unit of power used by French engineers is called a Force de cheval. It is equivalent to 75 kilogramme-metres per second or to about 542 foot-pounds per second. Thus it is a little less than one horse-power.

# EXAMPLES.

**1.** *Find the work done in dragging a heavy* **body up an** *inclined plane.*

Let $AB$ be the inclined plane, $BD$ the perpendicular drawn **from** $B$ on the horizontal plane through $A$, and let the angle $BAD = a$.

If $W$ be the weight of the body the resolved part of the weight down the plane is $W \sin a$, and the resolved part perpendicular to the plane is $W \cos a$.

Hence the resistance due to friction during the motion is $\mu W \cos a$, where $\mu$ is the coefficient of friction. Hence the total force to be overcome is $W(\sin a + \mu \cos a)$.

∴ work done in dragging the body up a distance $AB$

$= W(\sin a + \mu \cos a) \cdot AB = W \cdot DB + \mu W \cdot AD$

= work done in raising the weight through a vertical distance $DB$, together with the **work** done in **dragging it a** distance equal to $AD$ along **a** horizontal plane **of the same roughness as** the inclined plane.

**2.** *Find the work done in stretching an elastic string.*

Let $OA\ (= a)$ be the unstretched length of the string, and let us find the work done in stretching it from a length $OB\ (= b)$ to $OC\ (= c)$.

When the length is $OP$, the tension $= \dfrac{\lambda}{a}(OP - a) = \dfrac{\lambda}{a} PA$.

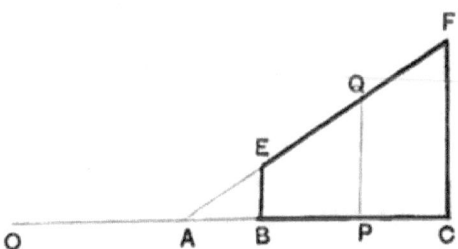

Erect at $P$ a perpendicular $PQ$ to represent this tension.

Then $\dfrac{PQ}{PA} = \dfrac{\lambda}{a}$; hence the angle $PAQ$ is constant, and **therefore** $Q$ always lies on a straight line through $A$. If this meet the perpendiculars through $B$, $C$ in $E$, $F$ then the required work is, as in Art. 58, represented by the area $BEFC$, and hence is $\frac{1}{2} BC \times (BE + CF)$ **or is equal to the**

*Extension produced multiplied by the mean of the* **initial and final** *tensions.*

# EXAMPLES.

**3.** *A train of mass* 100 *tons is ascending uniformly an incline of* 1 *in* 280 *and the resistance due to friction etc. is equal to* 16 *lbs. per ton; if the engine be of* 200 *H.-P. and be working at full power, find the rate at which the train is going.*

The resistance due to friction etc. is equal to the weight of 1600 pounds, and the resolved part of the weight of the train down the incline is equal to the weight of $\frac{1}{280}$ of 100 tons, or to the weight of 800 lbs.; hence the total force to impede the motion is equal to the weight of 2400 lbs.

Let $v$ be the velocity of the train in feet per second. Then the work done by the engine is that done in dragging a force equal to the weight of 2400 lbs. through $v$ feet per second and is equivalent to $2400v$ foot-pounds per second.

But the total work which the engine can do is $\dfrac{200 \times 33000}{60}$ or 110,000 foot-pounds per second.

Hence $$2400v = 110000,$$
or $$v = \frac{1100}{24},$$
and hence the velocity of the train is $31\frac{1}{4}$ miles per hour.

**4.** *A train of mass* 80 *tons is, at a certain instant, moving at the rate of* 30 *miles per hour up an incline of* 1 *in* 100, *the resistances due to friction etc. being equal to* 14 *lbs. per ton; if the engine be of* 250 *horse-power and be working at full power, find the acceleration of the train at that instant.*

The train is moving at the rate of 44 feet per second, and the engine is doing $250 \times 550$ foot-pounds per second, and is therefore exerting a force equal to the weight of $\dfrac{250 \times 550}{44}$ or 3125 pounds.

Now the force down the plane due to the weight of the train and the friction $= 80\,[2240 \div 100 + 14]$ pounds weight

$\quad = 2912$ pounds weight.

$\therefore$ the force to accelerate the train $= 213$ pounds weight

$$= 213 \times 32 \text{ poundals.}$$

$\therefore$ acceleration $= \dfrac{213 \times 32}{80 \times 2240}$ ft.-sec. units $= \dfrac{213}{5600}$ ft.-sec. units.

EXAMPLES. 91

5. Find the H.-P. of an engine which will travel at the rate of 25 miles per hour up an incline of 1 in 100, the mass of the engine and load being 10 tons and the resistances being 10 lbs. per ton.

*Ans.* 21⅔.

6. A train of mass 200 tons, including the engine, is drawn up an incline of 3 in 500 at the rate of 40 miles per hour by an engine of 600 H.-P.; find the resistance per ton due to friction etc.

*Ans.* 14·685 lbs. weight.

7. A weight of 10 tons is dragged in half-an-hour a length of 330 feet up a rough plane inclined at an angle of 30° to the horizon; the coefficient of friction being $\frac{1}{\sqrt{3}}$ find the work expended, and the H.-P. of an engine by which it will be done.

*Ans.* 7,392,000 ft.-lbs.; 7·4̇6̇ H.-P.

8. What is the horse-power of an engine which keeps a train of 100 tons going at the rate of 40 miles per hour against a resistance of 2000 lbs.?

*Ans.* 213⅓ H.-P.

9. The least H.-P. of an engine which is taking a train of 120 tons up an incline of 1 in 224 at the rate of 30 miles per hour is 336, supposing a resistance of 25 lbs. per ton on the level at this speed.

10. Shew that the work done in raising a given mass $M$ from one position to another is $Mgh$, where $h$ is the distance through which the centre of inertia of the mass has been raised vertically.

11. In how many hours would an engine of 18 H.-P. empty a vertical shaft full of water, if the diameter of the shaft be 9 feet, the depth 420 feet, and the mass of a cubic foot of water 62½ lbs.?

*Ans.* 9$\frac{37}{77}$ hours.

12. An elastic string is gradually stretched by an increasing force; when the string has been lengthened 2 inches the tension is found to be equal to the weight of 10 lbs. What is the work that has been done in stretching the string?

*Ans.* ⅚$g$ foot-poundals.

**92. Energy. Def.** *The Energy of a body is its capacity for doing work and is of two kinds, Kinetic and Potential.*

*The* **Kinetic Energy** *of a body is the energy which it possesses by virtue of its motion and is measured by the amount of work that the body can perform against the impressed forces before its velocity is destroyed.*

A falling body, a swinging pendulum, and a cannon ball in motion all possess kinetic energy.

Consider the case of a particle, of mass $m$, moving with velocity $u$, and let us find the work done by it before it comes to rest.

Suppose it brought to rest by a constant force $F$ resisting its motion, which produces in it an acceleration $-f$ given by $F = mf$.

Let $x$ be the space described by the body before it comes to rest so that
$$0 = u^2 + 2(-f) \cdot x.$$
$$\therefore fx = \tfrac{1}{2}u^2.$$

Hence the kinetic energy of the body
= work done by it before it comes to rest
$= Fx = mfx = \tfrac{1}{2}mu^2.$

Hence the kinetic energy is equal to the product of the mass of the body and one half the square of its velocity.

The definition given above therefore coincides with that given in Art. 68.

**93.** *The* **Potential Energy** *of a body is the work it can do by means of its position in passing from its present configuration to some standard configuration* (*usually called its zero position*).

A bent spring has potential energy, viz. the **work it can do in recovering its position**. **A body raised to a** height above the ground has potential energy, viz. the work its weight **can** do as it falls to the earth's surface.

The following example is important:

94. *A particle of mass* m *falls from rest* at a *height* h *above the* **ground**; *to shew that the sum* **of** *its potential and kinetic* **energies** *is constant throughout the motion.*

Let $H$ be the point from which the particle starts and $O$ the point where it reaches the ground.

Let $v$ be its velocity when it has fallen through a distance $HP\,(=x)$, so that $v^2 = 2gx$.

∴ kinetic energy there $= \tfrac{1}{2}mv^2 = mgx$.

Also its potential energy at $P$

$=$ the work its weight can do as it falls from $P$ to $O$

$= mg \cdot PO = mg\,(h-x)$.

∴ the sum of its kinetic and potential energies at $P$

$$= mgh.$$

But its potential energy when at $H$ is $mgh$ and its kinetic energy there is zero.

Hence the sum of the potential and kinetic energies is the same at $P$ as at $H$; and since $P$ is *any* point it follows that the sum of these two quantities is the same throughout the motion.

95. The above is an extremely simple example of the principle of the **Conservation of Energy**, which may be stated as follows:

*If a body* **or system of bodies** *be in motion under a conservative system of forces*, **the sum of** *its kinetic and potential energies is constant.*

*A conservative system of forces is one which depends on the position or configuration* **only** *of the system of bodies, and not on the velocity or direction of motion of the bodies.*

Thus from a conservative system are excluded forces of the nature of friction, or forces such as the resistance of the air which varies as some power of the velocity of the body. Friction is excluded because if the direction of motion of the body be reversed, the direction of the friction is reversed also.

Referring to the case of a particle sliding down a rough plane of length $l$ (Art. 84, *Ex.* 1) we see that the kinetic energy of the particle on reaching the ground is $\frac{1}{2}m\left[2gl\left(\sin\alpha - \mu\cos\alpha\right)\right]$, or $mgl\sin\alpha - mgl\mu\cos\alpha$. Also the potential energy there is zero, so that the sum of the kinetic and potential energies at the foot of the plane is

$$mgl\sin\alpha - \mu mgl\cos\alpha.$$

But the potential energy of the particle when at the top of the plane is $mg \cdot l\sin\alpha$, so that the total loss of visible mechanical energy of the particle in sliding from the top to the bottom of the inclined plane is $\mu mgl\cos\alpha$. This energy has been transformed and appears chiefly in the form of heat, partly in the moving body, and partly in the plane.

96.  **Theorem.** *To shew that the change of kinetic energy per unit of space is equal to the acting force.*

If a force $F$ acting on a particle of mass $m$ change the velocity from $u$ to $v$ in time $t$ whilst the particle moves through a space $s$, we have $v^2 - u^2 = 2fs$, where $f$ is the acceleration produced.

$$\therefore \frac{\frac{1}{2}mv^2 - \frac{1}{2}mu^2}{s} = mf = F\ldots\ldots\ldots\ldots(1).$$

This equation proves the proposition when the force is constant.

When the force is variable, the same proof will hold if we take $t$ so small that the force $F$ does not sensibly alter during that interval.

*Cor.* It follows from equation (1) that the change in the kinetic energy of a particle is equal to the work done on it.

Ex. 1. Find the kinetic energy in ergs of a cannon ball of 10,000 grammes discharged with a velocity of 5000 centimetres per second.

*Ans.* $1{\cdot}25 \times 10^{11}$ ergs.

Ex. 2. A cannon ball, of mass 5000 grammes, is discharged with a velocity of 500 metres per second. Find its kinetic energy in ergs, and, if the cannon be free to move and have a mass of 100 kilogrammes, find the energy of the recoil.

*Ans.* $6{\cdot}25 \times 10^{12}$ ergs ; $3{\cdot}125 \times 10^{11}$ ergs.

Ex. 3. A bullet of mass 50 grammes is fired into a target with a velocity of 500 metres per second. The target is of mass one kilogramme and is free to move; find the loss of energy by the impact in kilogrammetres.

*Ans.* $5952\frac{8}{21}$ kilogrammetres.

Ex. 4. Find the H.-P. of a steam gun which projects 100 four-pound shots in a minute with a velocity of 1200 feet per second.

*Ans.* $272\frac{8}{11}$.

## 97. Graphic Method. Force-Space Curve.

Since the kinetic energy acquired by a particle in a given time is equal to the work done on the particle in that time, it follows that if in the figure of Art. 58 the abscissae represent the distances described by a particle, and the ordinates represent the corresponding forces acting on the particle, then the kinetic energy generated is represented by the number of units of area in $OACB$.

**Ex. 1.** If a particle be moving in a straight line under a force to a fixed point in the straight line, which varies as the distance of the particle from the fixed point, find the kinetic energy of the particle when a given distance has been described from rest.

**Ex. 2.** If in the last example, the force vary as the square of the distance of the particle from the fixed point, find the kinetic energy acquired.

*Motion of the centre of inertia of a system of particles.*

**98. Theorem.** *If the velocities at any instant of any number of masses* $m_1$, $m_2$, $m_3$ ...... *parallel to any line fixed in space be* $u_1$, $u_2$, $u_3$ ...... *then the velocity parallel to that line of the centre of inertia of these masses at that instant is*

$$\frac{m_1 u_1 + m_2 u_2 + \ldots}{m_1 + m_2 + \ldots}.$$

At the instant under consideration, let $x_1, x_2, x_3$ ...... be the distances of the given masses measured along this fixed line from a fixed point in it and let $\bar{x}$ be the distance of their centre of inertia.

Then (as is shewn in any treatise on Statics)

$$\bar{x} = \frac{m_1 x_1 + m_2 x_2 + \ldots}{m_1 + m_2 + \ldots}.$$

Let $x_1'$, $x_2'$ ...... be the corresponding distances of these masses at the end of a small time $t$ and $\bar{x}'$ the corresponding distance of their centre of inertia. Then we have

$$x_1' = x_1 + u_1 t, \quad x_2' = x_2 + u_2 t, \quad \ldots\ldots\ldots\ldots$$

Also $\qquad \bar{x}' = \dfrac{m_1 x_1' + m_2 x_2' + \ldots}{m_1 + m_2 + \ldots},$

$$\therefore \bar{x}' - \bar{x} = \frac{m_1 (x_1' - x_1) + m_2 (x_2' - x_2) + \ldots}{m_1 + m_2 + \ldots}$$

$$= \frac{m_1 u_1 t + m_2 u_2 t + \ldots}{m_1 + m_2 + \ldots}.$$

But if $\bar{u}$ be the velocity of the centre of inertia parallel to the fixed line we have $\bar{x}' = \bar{x} + \bar{u}t$.

$$\therefore \bar{u} = \frac{\bar{x}' - \bar{x}}{t} = \frac{m_1 u_1 + m_2 u_2 + \ldots}{m_1 + m_2 + \ldots}.$$

Hence the velocity of the centre of inertia of a system of particles in any given direction is equal to the sum of the momenta of the particles in that direction divided by the sum of the masses of the particles.

*Cor.* If a system of particles be in motion in a plane, and their velocities and directions of motion be known, we can by resolving these velocities parallel to two fixed lines and applying the preceding proposition find the motion of their centre of inertia.

So if we have a system of particles in motion in space we must resolve the velocities along three fixed lines not lying in one plane and proceed as before.

99. **Theorem.** *If the accelerations at any instant of any number of masses* $m_1$, $m_2$, $m_3$ ...... *parallel to any line fixed in space be* $f_1$, $f_2$, $f_3$ ...... *then the acceleration of the centre of inertia of these masses parallel to this line is*

$$\frac{m_1 f_1 + m_2 f_2 + \ldots}{m_1 + m_2 + \ldots}.$$

The proof of this proposition is similar to that of the last article. We have only to change $x_1$, $u_1$, $x_1'$, $u_1'$ into $u_1$, $f_1$, $u_1'$, $f_1'$, and make similar changes for the other particles.

100. **Theorem.** *The motion of the centre of inertia of a system of particles is unaffected by any mutual action between the particles of the system.*

For consider any two particles $m_1$, $m_2$ of the system and the stress between them.

[This stress may be of the nature of an impact, as when two billiard balls impinge, or of the nature of an attraction as when the earth and a falling body attract one another.]

Then, by the 3rd law, the action on $m_1$ at any instant is equal and opposite to that on $m_2$; hence the impulse of the action on $m_1$ is equal and opposite to the impulse of the action on $m_2$.

It follows that the change in the momentum of $m_1$ due to this mutual action

$= -$ [change in the momentum of $m_2$ due to this action].

Hence the sum of the changes in their momenta due to this action is zero, or the sum of the momenta of $m_1$, $m_2$ is not altered by their mutual action.

Hence the sum of the momenta of the two masses resolved along any fixed line is unaffected by the stress between them.

The same result holds for every such pair of particles. But the velocity of the centre of inertia in any direction

$$= \frac{\text{momentum of the particles in that direction}}{\text{sum of their masses}}.$$

Hence the velocity of the centre of inertia is unaltered by the stresses of the system.

*Cor.* It follows that the first law of motion might be enunciated as follows; *The centre of inertia of any system of particles will continue at rest, or in uniform motion in a straight line, except it be compelled by external forces to change that state of rest or uniform motion.*

**101. Theorem.** *The motion of the centre of inertia of any system* of particles is the same as *if all the masses of the system were collected at the centre of inertia,* and all the external *forces of the system applied there parallel to their original directions.*

Let the masses of the different particles of the system be $m_1, m_2 \ldots$ and let the components of the external forces acting on them parallel to any line fixed in space be $P_1, P_2 \ldots$

Now the rate of change of the momentum of each particle is equal to the force acting on it parallel to the fixed direction.

∴ the rate of change of the sum of the momenta of the particles is equal to the sum of the forces acting on the particles parallel to the fixed direction.

But these latter forces consist of the external forces of the system together with the mutual actions of the system, and these latter are in equilibrium *inter se*.

Hence
$$m_1 f_1 + m_2 f_2 + \ldots = P_1 + P_2 + \ldots \ldots \ldots (i),$$

where $f_1, f_2 \ldots$ are the accelerations of the particles in the given direction.

Let $\bar{f}$ be the acceleration of the centre of inertia in this direction.

$$\therefore \bar{f} = \frac{m_1 f_1 + m_2 f_2 + \ldots}{m_1 + m_2 + \ldots} = \frac{P_1 + P_2 + \ldots}{m_1 + m_2 + \ldots},$$

and this latter acceleration is the same as would be produced by a force equal to $P_1 + P_2 + \ldots$ acting on a mass equal to the sum of the masses.

The same would be true for any other direction, and

hence the theorem follows by compounding the accelerations and the forces.

*Cor.* I. The equation (i) of the preceding article states that the rate of change of the momentum of any system in a given direction is equal to the components of the external forces in that direction. Hence similarly as in Art. 85 it follows that

*The change of momentum of any system in any given direction during any interval of time is measured by the sum of the impulses of the external forces in the given direction during that interval.*

*Hence, also, if the sum of the external forces acting on any system resolved in any given direction always vanish, the total momentum of the system in that direction remains the same throughout the motion.*

This latter theorem is often referred to as the principle of the **Conservation of Momentum**.

*Cor.* II. The motion of the centre of inertia of a heavy chain, falling freely, is the same as that of a particle falling freely.

### EXAMPLES.

1. Two masses $m_1$, $m_2$ are connected by a light string as in Art. 83; to find the acceleration of the centre of inertia of the system.

The acceleration of the mass $m_1$ is $\dfrac{m_1 - m_2}{m_1 + m_2} g$ vertically downwards, and that of $m_2$ is the same in the opposite direction.

Here then $f_1 = -f_2 = \dfrac{m_1 - m_2}{m_1 + m_2} g$,

$\therefore$ acceleration of the centre of inertia $= \dfrac{m_1 f_1 + m_2 f_2}{m_1 + m_2} = \left(\dfrac{m_1 - m_2}{m_1 + m_2}\right)^2 g$.

2. Two masses move at a uniform rate along two straight lines which meet and are inclined at a given angle; shew that their centre of inertia describes a straight line with uniform velocity.

# EXAMPLES. 101

3. Two masses $m_1$, $m_2$ rest on a rough double inclined plane, being connected by a light string passing over a small smooth pulley at the vertex of the inclined plane, the inclination of whose faces to the horizon are respectively 60° and 30°. If the angle of friction be 30°, and motion be allowed to ensue, find the ratio of $m_1$ to $m_2$ when the vertical acceleration of their centre of inertia is one quarter the acceleration of a freely falling particle.

*Ans.* $m_1 : m_2 :: 3 + 2\sqrt{3} : 1$.

4. Of three equal particles which start from the highest point of a vertical circle, one drops down a vertical diameter, and the others slide down chords of 60° and 120° on the same side of the diameter; shew that the acceleration of the centre of inertia is $\frac{1}{6}\sqrt{19}g$ down a chord of $\cos^{-1}\left(\dfrac{-13}{19}\right)$.

5. Two masses are connected by a string which passes over the top of two inclined planes of equal height placed back to back. Shew that the path of their centre of inertia is that line which joins them when they are in such a position that the parts of the string on the two planes are to one another as the masses at their extremities.

## MISCELLANEOUS EXAMPLES.

1. *Three inches of rain fall in a certain district in* 12 *hours. Assuming that the drops fall from a height of a quarter of a mile, find the pressure on the ground per square mile of the district due to the rain during the storm, the mass of a cubic foot of water being* 1000 *ounces.*

The amount of rain that falls on a square foot during the storm is $\frac{1}{4}$ of a cubic foot, and its mass is 250 ounces.

∴ the mass that falls per second is $\dfrac{250}{16} \times \dfrac{1}{12 \cdot 60 \cdot 60}$ or $\dfrac{5}{144 \times 96}$ lbs.

The velocity of each raindrop on touching the ground is

$$\sqrt{2 \times g \times 440 \times 3} \text{ or } 16\sqrt{330} \text{ ft. per second.}$$

∴ the momentum that is destroyed per second is

$$\dfrac{5}{144 \times 96} \times 16\sqrt{330}, \text{ or } \dfrac{5\sqrt{330}}{864} \text{ units of momentum.}$$

But the number of units of momentum destroyed per second is equal to the number of poundals in the acting force.

∴ pressure on the ground per square foot $= \dfrac{5\sqrt{330}}{864}$ poundals.

∴ pressure per square mile = weight of $9 \times 4840 \times 640 \times \dfrac{5\sqrt{330}}{32 \times 864}$ lbs.

= weight of 41 tons approximately.

**2.** *The scale-pans of a balance are each of mass* M, *and in them are placed masses* $M_1$ *and* $M_2$; *to shew that the pressures on the pans during the motion are*

$$\dfrac{2M_1(M+M_2)}{M_1+M_2+2M}\cdot g \quad\text{and}\quad \dfrac{2M_2(M+M_1)}{M_1+M_2+2M}\cdot g$$

*respectively.*

Let $f$ be the common acceleration of the system and suppose $M_1 > M_2$. Then as in Art. 83 we have

$$f = (M_1 - M_2)\,g/(2M + M_1 + M_2).$$

Let $P$ be the pressure between $M_1$ and the scale-pan on which it rests; then the force on the mass $M_1$, *considered as a separate body*, is $M_1 g - P$. Also its acceleration is $f$.

Hence $\quad M_1 g - P = M_1 f$,

$$\therefore\; P = M_1(g - f) = 2M_1(M + M_2)\,g/(2M + M_1 + M_2).$$

**3.** *A mass* P *after falling freely through a distance of* a *feet begins to raise a mass* Q, *greater than itself, being connected with it by means of a fine string passing over a smooth pulley. Find the resulting motion.*

Let $v$ be the velocity of the mass $P$ just before the string becomes tight so that $v = \sqrt{2ga}$.

When the string becomes tight a jerk is immediately felt on $P$ and similarly on $Q$. This jerk, consisting of a very large force acting for an exceedingly short time, is of the nature of an impulsive force, and its effect is measured by the change of momentum produced. Immediately after the string becomes tight, the masses are moving with a common velocity, $V$.

Then the impulse of the force is measured by the change in the momentum of $P$, and also by the change in that of $Q$, and these are respectively $P(v - V)$ and $QV$.

Hence $\qquad P(v - V) = QV$,

$$\therefore\; V = \dfrac{P}{P + Q}\,v,$$

and the impulse of the jerk $= QV = \dfrac{PQ}{P+Q}\,v.$

When the string becomes tight the acceleration of $Q$ is by Art. 83, $\frac{Q-P}{Q+P} g$ in a downward direction, and hence $Q$ ascends a distance $x$ given by $\left(\frac{P}{P+Q} v\right)^2 = 2 \frac{Q-P}{Q+P} gx$, so that $x = a \frac{P^2}{Q^2 - P^2}$, and then turns back and moves in the opposite direction.

4. *Two balls* A, B *of masses* $m_1$, $m_2$ *are connected by an inextensible string and placed on a smooth table; one of them*, B, *is set in motion with a given velocity in a given direction, to find the motion of each immediately after the string has become tight.*

Immediately before the string becomes tight let the velocity of $B$ be $u$ at an angle $a$ with the string. Now the action of the string during the impact is along $AB$ only, and hence the velocity $u \sin a$ of $B$ perpendicular to $AB$ is unaltered. Also since the string is inextensible the velocities of $A$, $B$ along the line $AB$ immediately after the string has become tight must be the same, $V$. Then immediately after the string is tight, the momentum of the system along $AB$ is $(m_1 + m_2) V$, and this must, as in Art. 86, be equal to $m_2 u \cos a$.

Hence the instant after the string becomes tight the velocities of $B$ along and perpendicular to $AB$ are $m_2 u \cos a/(m_1 + m_2)$ and $u \sin a$.

## EXAMPLES. CHAPTER II.

1. Two scale-pans, each of mass 2 ounces, are suspended by a weightless string over a smooth pulley; a mass of 10 ounces is placed in the one, and 4 ounces in the other. Find the tension of the string and the pressures on the scale-pans.

2. A shot of mass $m$ is fired from a gun of mass $M$ with velocity $u$ relative to the gun; shew that the actual velocities of the shot and gun are $\frac{Mu}{M+m}$ and $\frac{mu}{M+m}$ respectively.

3. If a shot be fired from a gun, shew that, neglecting the mass of the powder, the work done on the shot and gun respectively are inversely proportional to their masses.

# EXAMPLES ON CHAPTER II.

4. A shot, whose mass is 800 pounds, is discharged from an 81-ton gun with a velocity of 1400 feet per second; find the steady pressure which acting on the gun would stop it after a recoil of five feet.

5. A gun is mounted on a gun-carriage movable on a smooth horizontal plane and the gun is elevated at an angle $a$ to the horizon; a shot is fired and leaves the gun in a direction inclined at an angle $\theta$ to the horizon; if the mass of the gun and its carriage be $n$ times that of the shot, shew that

$$\tan \theta = \left(1 + \frac{1}{n}\right) \tan a.$$

6. Find the energy per second of a waterfall 30 yards high and a quarter of a mile broad, where the water is 20 feet deep and has a velocity of $7\frac{1}{2}$ miles per hour when it arrives at the fall. The mass of the water is 1024 ounces per cubic foot.

7. A steam hammer of mass 20 tons falls vertically through 5 feet, being pressed downwards by steam pressure equal to the weight of 30 tons; what velocity will it acquire, and how many foot-pounds of work will it do before coming to rest?

8. Find in horse-power the rate at which a locomotive must be able to work in order to get up a velocity of 20 miles per hour on a level line in a train of 60 tons in 3 minutes after starting, the resistance to motion being taken at 10 pounds per ton.

9. Shew that a train going at the rate of 30 miles per hour will be brought to rest in about 84 yards by continuous brakes, if they press on the wheels with a force equal to $\frac{3}{4}$ of the weight of the train, the coefficient of friction being ·16.

EXAMPLES ON CHAPTER II.  105

10. A train of 150 tons moving with a velocity of 50 miles per hour has its steam shut off and the brakes applied and is stopped in 363 yards. Supposing the resistance to its motion to be uniform, find its value and find also the mechanical work done by it measured in foot-pounds.

11. The driver of an express train which is running at the rate of 70 miles per hour observes 660 yards ahead the tail-lights of a goods train at rest. The brakes of the express train, which exert a constant retardation, can pull it up in 500 yards when running with steam off at 50 miles per hour, and the goods train can in one minute get up a velocity of 30 miles per hour. If the express shuts off steam and puts on its brakes and the goods train puts on full steam ahead, find the respective velocities just before collision.

12. On a certain day half an inch of rain fell in 3 hours; assuming that the drops were indefinitely small and that the terminal velocity was 10 feet per second, find the impulsive pressure in tons per square mile consequent on their being reduced to rest, assuming that the mass of a cubic foot of water is 1000 ounces and that the rain was uniform and continuous.

13. Find the pressure in poundals per acre due to the impact of a fall of rain of 3 inches in 24 hours, supposing the rain to have a velocity due to falling freely 400 feet.

14. A cage of mass 2 tons is lifted in 2 minutes from the bottom of a pit a quarter of a mile deep, by means of an engine which exerts a constant tractive force, but the steam is shut off during the ascent so that the cage just comes to rest at the surface. Find the smallest possible horse-power of the engine.

15. A train of mass 200 tons moving with a velocity of $21\tfrac{9}{11}$ miles per hour comes to the top of an incline of 1 in 150, and runs down with steam turned off for 1800 yards when it comes to the foot of a similar incline; find the least horse-power which will carry it to the top in one minute, supposing the resistance due to friction etc., to be equal to 11·2 lbs. per ton.

16. A train of mass 200 tons is ascending an incline of 1 in 100 at the rate of 30 miles per hour, the resistance of the rails being equal to the weight of 8 lbs. per ton. The steam being shut off and the brakes applied the train is stopped in a quarter of a mile. Find the weight of the brake-van, the coefficient of sliding friction of iron on iron being $\tfrac{1}{8}$.

17. Assuming that the resistances of all kinds to the motion of a train on the level are equal to 6 lbs. per ton of the total weight of the train when it is moving with less speed than 10 miles per hour, with an addition of $\tfrac{1}{8}$ lb. per ton for every mile of speed above 10, and that the total weight of the train and the engine is 200 tons, and that the maximum speed attainable is 40 miles per hour, find the time from rest of reaching a velocity of 10 miles per hour, the engine always exerting its full force.

18. A carriage is slipped from an express train going at full speed at a distance $l$ from a station; shew that, if the carriage come to rest at the station itself, the rest of the train will then be at a distance $\dfrac{M}{M-m}l$ beyond the station, $M$ and $m$ being the masses of the whole train and of the carriage slipped, and the pull exerted by the engine being constant.

19. From King's Cross to Grantham is 105 miles and there are 27 intermediate stations. A parliamentary train stops at all stations and the average resistance to its motion when the brake is off is $\tfrac{1}{280}$th of its weight whilst the resistance to an

express train which runs the distance without stopping is $\frac{1}{224}$th of its weight, but, when the brake is on, the resistance in each case is $\frac{1}{28}$th of its weight. Supposing the brake to be always applied when the speed has been reduced to 30 miles per hour and not before, find which train is most expensive and by how much per cent.

20. A train of mass $m$ moves against a uniform retardation equal to $\mu$ times its weight, and starts from rest moving with uniform acceleration until the steam is shut off, and arrives at the next station distant $a$ from the starting point in time $t$. Shew that the greatest horse-power exerted by the engine is
$C\dfrac{2m\mu^2 g^2 at}{\mu g t^2 - 2a}$, where $C$ is a constant depending on the units employed.

21. The resistance $R$ due to the air and friction to a train moving at the rate of $V$ miles per hour ($V$ being $> 10$) is given in pounds by the formula $R = [6 + \frac{1}{3}(V-10)][T + 2E]$ where $E$ is the weight of the engine and $T$ that of the train in tons. If $V$ be $< 10$ then $R = 6(T + 2E)$. Assuming this formula, and that the force exerted by the engine is constant, that the weight of the engine is 40 tons and that of the train 160 tons and that 40 miles per hour is full speed for the train, find its velocity at the end of 1 minute 24 seconds from the start.

22. An engine of horse-power $H$ draws a train of $M$ tons up an incline of 1 in $m$, the resistance to motion apart from gravity being equal to the weight of $n$ lbs. per ton. Shew that the maximum speed is

$$\dfrac{550Hm}{M(2240 + nm)} \text{ feet per second.}$$

23. Two equal balls are let fall at the same instant, one from the ridge of a house down the slates and the other from the eaves. The balls roll so that when one touches the ground

the other is at the edge of the eaves, and the perpendicular to the earth will pass through both balls; find the locus of the centre of inertia of the balls during the motion.

24. Two equal masses, $A$ and $B$, are connected by an inelastic thread, 3 feet long, and are laid close together on a smooth horizontal table $3\frac{1}{2}$ feet from its nearest edge; $B$ is also connected by a stretched inelastic thread with an equal mass $C$ hanging over the edge. Find the velocity of the masses when $A$ begins to move and also when $B$ arrives at the edge of the table.

25. Two masses, each equal to $W$, are hung over a smooth pulley, being connected by a string sixteen feet long, and are arranged so that either on approaching to within two feet from the pulley lifts a load $\dfrac{2W}{17}$. The motion is started with one weight and its load at the pulley; shew that the motion repeats itself every fifteen seconds.

26. Two masses $P$, $Q$ connected by a string passing over a smooth pulley are both held at a distance $c$ above a fixed horizontal plane and motion ensues. If the plane destroy the momentum of the heavier mass $P$ as it hits the plane, shew that the system comes to rest in time

$$.3\left\{\frac{P+Q}{P-Q}\cdot\frac{2c}{g}\right\}^{\frac{1}{2}}.$$

27. A body of mass $m$ lbs. is attached to a weightless inextensible string which passes over a smooth pulley, and is attached at the other end to a body of mass $m'$ lbs. lying on a table, the line joining the position of $m'$ to the pulley being inclined at an angle $a$ to the vertical. If the string be at first slack, and become stretched when the body has fallen through one foot, find the impulse of the tension and the velocity communicated to $m'$.

28. A mass $M$ after falling freely through $a$ feet begins to raise a mass $m$ greater than itself and connected with it by means of an inextensible string passing over a fixed pulley. Shew that $m$ will have returned to its original position at the end of time

$$\frac{2M}{m-M}\sqrt{\frac{2a}{g}}.$$

29. An inelastic mass $M$ hanging freely draws another mass $M'$ up a rough inclined plane by means of a string over a pulley at the top of the plane. If $M$ start from the top of the plane and $M'$ from the bottom, find the velocity of $M'$ when $M$ strikes the ground and, if $\mu$ be $\dfrac{1}{\sqrt{3}}$ and the angle of inclination of the plane be 30°, find the subsequent motion.

30. At the bottom of a mine 275 feet deep there is an iron cage containing coal weighing 14 cwt., the cage itself weighing 4 cwt. 109 lbs. and the wire rope that draws it 6 lbs. per yard. Find the work done when the load has been lifted to the surface, and the horse-power of the engine required to do that amount of work in forty seconds.

31. A man of 12 stone ascends a mountain 11000 feet high in 7 hours, and the difficulties in his way are equivalent to carrying a weight of 3 stone; one of Watt's horses could pull him up the same height without impediments in 56 minutes; shew that the horse does as much work as 6 such men in the same time.

32. A blacksmith wielding a 14-lb. sledge strikes an iron bar 25 times per minute and brings the sledge to rest upon the bar after each blow. If the velocity of the sledge on striking the iron be 32 feet per second, compare the rate at which he is working with a horse-power.

33. A train of 120 tons is to be taken from one station to another, a mile off, up an incline of 1 in 80 in four minutes without using brakes. Shew that neglecting passive resistances the engine must exert a pull before the steam is turned off equal to the weight of about 6203 pounds.

34. Find the loss of time and the extra expenditure of work due to crossing a pass by a railway having an incline of one in $m$ up and one in $n$ down, instead of going through a level tunnel of length $l$ through the pass, supposing that $V$ is the maximum speed allowed on the line and that the speed drops from $V$ at the bottom to $v$ at the summit of the pass.

35. Shew that the loss of time in going from $A$ to $C$, two points on a railway at the same level 8 miles apart, due to an incline of 1 in 100 from $A$ up to $B$ and an incline of 1 in 300 from $B$ down to $C$, instead of going from $A$ to $C$ at a uniform velocity of 45 miles per hour, is about 2 minutes 20 seconds.

It is supposed that with full steam on the velocity drops from 45 miles an hour at $A$ to 15 at the summit $B$, and that in descending the incline from $B$ to $C$ full steam is again kept on till the velocity has again reached 45 miles per hour, after which the velocity is kept uniform by partly shutting off steam; and prove that this happens at a point $Q$ distant from $B$ 1 mile 892 yards.

36. If the mass of the train in the previous question be 200 tons, and the resistance due to friction etc. be 14 pounds per ton, then the pull of the engine from $A$ to $Q$ is $2\frac{5}{48}$ tons, and from $Q$ to $C$, $\frac{7}{12}$ of a ton; find also the extra expenditure of work due to the inclines.

37. In an Attwood's machine, if the string can only bear a strain of one-fourth of the sum of the weights at the two extremities, shew that the larger weight cannot be much less than 6 times the smaller and that the least possible acceleration is $\dfrac{g}{\sqrt{2}}$.

EXAMPLES ON CHAPTER II.

38. The locus of points from which the times down equally rough inclined planes to a fixed point vary as the length of the planes is a right circular cone.

39. Particles slide down rough chords of an ellipse from the upper extremity of the major axis which is vertical. The chord drawn to the extremity of the minor axis will be the chord of quickest descent to the curve if the coefficient of friction be $\dfrac{e^2}{2\sqrt{1-e^2}}$, where $e$ is the eccentricity of the ellipse. If the coefficient of friction be greater than this value, then the major axis itself will be the chord of quickest descent.

40. Shew that points in a vertical plane from which a particle will slide down a rough chord to a fixed point in a given time lie on one or other of two fixed circles.

41. A pulley is suspended from a given point by a string; over the pulley passes another string to whose extremities are attached masses $M$ and $3M$. The masses are initially at rest; after five seconds have elapsed the string supporting the pulley is cut; find the distance of each mass from its initial position at the end of five seconds more.

42. A mass $m_1$ in descending draws up a mass $m_2$ by means of a string over a small pulley; if the latter after it has been in motion 10 seconds pick up another mass $m_3$, find the total distance described by it in twenty seconds from the time at which they start.

43. Two equal masses $P$, $Q$ connected by a string over a smooth pulley are moving with a common velocity, $P$ descending and $Q$ ascending. If $P$ be suddenly stopped and instantly let drop again find the time that elapses before the string is again tight.

EXAMPLES ON CHAPTER II.

44. A string over a pulley supports a mass of 5 lbs. on one side and masses of 2 and 3 lbs. on the other, the lower mass 2 lbs. being distant one foot from the other. The two-pound mass is suddenly raised to the same level as the other and kept from falling. Shew that the string will become taut in half a second, and that the whole system will then move with a uniform velocity of 3·2 ft. per sec.

45. A constant force acts on a particle at rest for a time $\frac{t}{n}$; it then ceases for time $\frac{t}{n}$, then acts for the same time, then ceases, and so on alternately. Find the space described in time $t$ and the limiting value when $n$ is infinite.

46. A heavy inelastic particle slides down an imperfectly rough plane inclined at an angle $\theta$ to the horizon; after describing a path, length $a$, it meets another equally rough plane inclined at an angle $\theta$ to the horizon up which it ascends describing a path of length $b$ before it comes to rest; shew that

$$b = a \cos^2 2\theta \cdot \frac{\sin(\theta - \epsilon)}{\sin(\theta + \epsilon)},$$

where $\epsilon$ is the angle of friction when the particle is in motion.

47. A cylinder, of height $h$ and diameter $d$, is standing on the horizontal seat of a railway carriage. If the train begin to move with acceleration $f$, shew that it will not remain undisturbed unless $f < \mu g$ and also $< \frac{dg}{h}$, where $\mu$ is the coefficient of friction.

48. Two equal balls $A$ and $B$ are at a distance $a$ apart. A force of impulse $I$ acts on $A$ in the direction $AB$ and a constant force $F$ on $B$ in the same direction produced. Shew that $A$ will not overtake $B$ unless $I^2 > 2aFm$, where $m$ is the mass of either ball.

EXAMPLES ON CHAPTER II.

49. Corn flows uniformly through an opening in a floor to a floor at a depth $h$ below; shew that the centre of inertia of the falling corn is at a distance $\dfrac{h}{3}$ below the opening.

50. Three particles are placed at the angular points of a triangle $ABC$, their masses being proportional to the opposite sides; shew that their centre of inertia is the centre of the inscribed circle of the triangle, and that if they move along the perpendiculars to the opposite sides with velocities respectively proportional to the cosines of the angles from which they started, the centre of inertia will move in a straight line towards the orthocentre.

51. A heavy particle is projected with velocity $v$ up a smooth inclined plane whose length and inclination are $l$ and $a$ respectively. At the same instant the plane starts off with uniform velocity $\dfrac{gl\tan a}{2v}$. Shew that the particle will reach the top in time $\dfrac{l}{v}$.

52. A heavy chain is drawn up by a given force $P$ which exceeds the weight, $W$, of the chain. Find its acceleration and the tension at any assigned point.

# CHAPTER III.

## THE LAWS OF MOTION (*continued*).

### MISCELLANEOUS EXAMPLES.

[In a first reading of the subject the student may, with advantage, omit this Chapter.]

**102.** THE following chapter consists of miscellaneous examples of a rather more difficult character than those given in Chapter II.

In some we have systems of masses whose motions are dependent on one another; in others the principles of the Conservation of Energy and Conservation of Momentum as enunciated in Articles **95** and **101** are exemplified.

**Ex. 1.** *A string passing over a smooth fixed* **pulley supports at its two ends two smooth** *moveable pulleys of masses* $m_1$, $m_2$ *respectively. Over each passes a string having masses* $m_1$, $m_2$ *at its ends; shew that the acceleration of each of the moveable pulleys is* $\dfrac{m_1^2 - m_2^2}{m_1^2 + 10 m_1 m_2 + m_2^2} g$, *and find the accelerations of the different masses, and the tensions of the strings.*

## MISCELLANEOUS EXAMPLES.

Let $A$, $B$ be the moveable pulleys of masses $m_1$, $m_2$. Let $C$, $E$ be the masses $m_1$ and $D$, $F$ the masses $m_2$. Let $T$ be the tension of the string round the fixed pulley, and $T_1$, $T_2$ the tensions of the strings round the pulleys $A$ and $B$, all three tensions being expressed in poundals.

Let $f$ be the acceleration of $A$ downwards or of $B$ upwards.

$$\therefore 2T_1 + m_1 g - T = m_1 f,$$

and $\qquad T - m_2 g - 2T_2 = m_2 f.$

Hence solving for $f$, $T$ we have

$$T(m_1 + m_2)$$
$$= 2T_1 m_2 + 2T_2 m_1 + 2m_1 m_2 g \ \ldots\ldots (1)$$

and $\qquad f = \dfrac{2(T_1 - T_2)}{m_1 + m_2} + \dfrac{m_1 - m_2}{m_1 + m_2} g \ \ldots\ldots (2)$

Also the **accelerations** of $C$ **and** $D$ downwards are respectively

$g - \dfrac{T_1}{m_1}$ and $g - \dfrac{T_1}{m_2}$.

But since in the course of the motion the string $AC$ lengthens as much as $AD$ shortens, therefore the motion of $C$ relative to $A$ must be equal and opposite to the motion of $D$ relative to $A$.

$\therefore$ acceleration of $C$ − acceleration of $A = -$ [**acc. of** $D$ − **acc. of** $A$], the accelerations being all measured downwards.

$$\therefore \text{ acc. of } C + \text{acc. of } D = 2 \cdot \text{ acc. of } A.$$

Hence $\qquad 2g - T_1 \left(\dfrac{1}{m_1} + \dfrac{1}{m_2}\right) = 2f,$

and $\qquad T_1 \left(\dfrac{1}{m_1} + \dfrac{1}{m_2}\right) = 2g - 2f \ \ldots\ldots\ldots\ldots (3).$

Similarly **acc.** of $E +$ **acc.** of $F = 2 \cdot$ acc. of $B = -2f.$

$$\therefore T_2 \left(\dfrac{1}{m_1} + \dfrac{1}{m_2}\right) = 2g + 2f \ \ldots\ldots\ldots (4).$$

From (3), (4) $\qquad (T_1 - T_2) \dfrac{m_1 + m_2}{m_1 m_2} = -4f.$

∴ substituting in (2),

$$f = -\frac{8m_1 m_2 f}{m_1+m_2} \cdot \frac{1}{m_1+m_2} + \frac{m_1-m_2}{m_1+m_2} g.$$

$$\therefore \quad f = \frac{m_1^2 - m_2^2}{m_1^2 + 10 m_1 m_2 + m_2^2} g.$$

Substituting this value of $f$ successively in equations (3) and (4) we obtain $T_1$ and $T_2$.

Also the acceleration of the mass $C$ downwards $= g - \dfrac{T_1}{m_1}$

$$= \frac{m_1-m_2}{m_1+m_2} g + \frac{2fm_2}{m_1+m_2}.$$

So the accelerations of the masses $D$, $E$, $F$ are obtained.

Ex. 2. *Two masses*, $M$ *and* $M_1$, *balance on a wheel and axle; if they be interchanged shew that the acceleration of* $M_1$ *is* $\dfrac{M_1 (M_1 - M)}{M_1^2 - M_1 M + M^2}$ g, *the mass of the wheel and axle being neglected.*

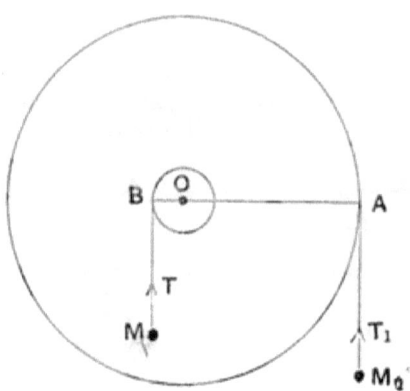

Let $OA$, $OB$ the radii of the wheel and axle respectively be $a$, $b$ so that when there is equilibrium we have

$$aM = bM_1 \quad\ldots\ldots\ldots\ldots\ldots\ldots\ldots\ldots\ldots\ldots\ldots(1).$$

When the masses are interchanged as in the above figure, let $T$, $T_1$ be the tensions of the strings, and $f$, $f_1$ the accelerations of $M$, $M_1$ respectively.

## WEDGE WITH MASSES ON ITS FACES.

Then
$$f_1 = (M_1 g - T_1)/M_1 \quad\quad\quad\quad\quad (2),$$
$$f = (T - Mg)/M \quad\quad\quad\quad\quad (3).$$

Now the resultant of $T$ and $T_1$ must be balanced by the reaction at $O$; for otherwise the forces on the machine will have a resultant, and this would cause an infinite acceleration in the machine, since its mass is negligible.

Hence
$$\frac{T}{T_1} = \frac{a}{b} = \frac{M_1}{M} \quad\quad\quad\quad\quad (4).$$

Again since the machine turns round its centre, the distances moved through by $M$, $M_1$ must be proportional to $b$ and $a$, and so also their accelerations.

Hence
$$\frac{f}{f_1} = \frac{b}{a} = \frac{M}{M_1} \quad\quad\quad\quad\quad (5).$$

**Hence from (2), (3), (5)** we have
$$\frac{M}{M_1} = \frac{T - Mg}{M} \cdot \frac{M_1}{M_1 g - T_1}.$$

$$M^2 (M_1 g - T_1) = M_1^2 (T - Mg) = M_1^2 \left[ \frac{M_1}{M} T_1 - Mg \right], \text{ by (4)}.$$

$$\therefore \quad T_1 = \frac{M^2 M_1}{M_1^2 - M_1 M + M^2} g.$$

Hence
$$f_1 = g - \frac{T_1}{M_1} g = \frac{M_1 (M_1 - M)}{M_1^2 - M_1 M + M^2} g.$$

**Ex. 3.** *A smooth isosceles wedge of mass* M *is placed on a smooth table and carries a smooth pulley at its summit, and two masses* $m_1$, $m_2$ *are attached to the ends of a string which passes over the pulley; shew that the acceleration with which the string passes over the pulley is*
$$\frac{m_1 - m_2}{m_1 + m_2} \cdot \frac{M + m_1 + m_2}{M + (m_1 + m_2) \sin^2 \alpha} \cdot g \sin \alpha,$$
*where* $\alpha$ *is the angle at the base of the wedge.*

*Find also the other circumstances of the motion.*

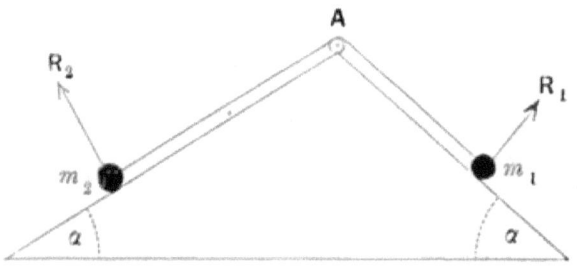

Let $f_1$, $f_2$ be the accelerations of $m_1$ along and perpendicular to its face of the wedge, $f_3$, $f_4$ those of $m_2$ similarly, $f_5$ that of the wedge horizontally, and $T$ the tension of the string. Then clearly

$$f_1 = (T - m_1 g \sin a)/m_1, \quad \ldots\ldots\ldots\ldots\ldots\ldots\ldots\ldots (1)$$

$$f_2 = (R_1 - m_1 g \cos a)/m_1, \quad \ldots\ldots\ldots\ldots\ldots\ldots\ldots (2)$$

$$f_3 = (T - m_2 g \sin a)/m_2, \quad \ldots\ldots\ldots\ldots\ldots\ldots\ldots (3)$$

$$f_4 = (R_2 - m_2 g \cos a)/m_2, \quad \ldots\ldots\ldots\ldots\ldots\ldots\ldots (4)$$

$$f_5 = (R_2 \sin a - R_1 \sin a)/M, \quad \ldots\ldots\ldots\ldots\ldots\ldots (5).$$

[In this last equation the horizontal components of the tension of the string vanish. Had the faces been inclined to the horizon at angles $a_1$, $a_2$, there would have been an additional term $T (\cos a_1 - \cos a_2)$ in the numerator of (5) due to the difference of the horizontal components of the tensions at the pulley.]

Also since the string is inextensible, the rate at which $m_2$ approaches $A$ is the same as that at which $m_1$ leaves $A$.

$$\therefore f_3 - f_5 \cos a = -f_1 - f_5 \cos a.$$

$$\therefore f_1 + f_3 = 0 \ldots\ldots\ldots\ldots\ldots\ldots\ldots\ldots\ldots\ldots (6).$$

Again the accelerations of $m_1$, $m_2$ respectively, perpendicular to the faces, must be the same as the accelerations of the wedge in the same directions; for the masses are always in contact with the wedge.

$$\therefore f_2 = f_5 \sin a \ldots\ldots\ldots\ldots\ldots\ldots\ldots\ldots (7)$$

$$-f_4 = f_5 \sin a \ldots\ldots\ldots\ldots\ldots\ldots\ldots\ldots (8).$$

From (1) (3) and (6) we obtain

$$-f_1 = f_3 = \frac{m_1 - m_2}{m_1 + m_2} g \sin a.$$

So from the remaining equations on solving

$$\frac{R_1}{m_1 (M + 2m_2 \sin^2 a)} = \frac{R_2}{m_2 (M + 2m_1 \sin^2 a)} = \frac{g \cos a}{M + (m_1 + m_2) \sin^2 a}.$$

$\therefore$ acceleration with which the string slips over the pulley

$$= f_3 - f_5 \cos a = f_3 - \cos a \sin a \frac{R_2 - R_1}{M}$$

$$= \frac{m_1 - m_2}{m_1 + m_2} \cdot \frac{M + m_1 + m_2}{M + (m_1 + m_2) \sin^2 a} \cdot g \sin a, \text{ on reduction.}$$

Also the acceleration of the wedge along the plane $= f_5 = \sin a \dfrac{R_2 - R_1}{M}$

$$= - \frac{m_1 - m_2}{M + (m_1 + m_2) \sin^2 a} g \sin a \cos a.$$

PERFORATION OF A PLATE BY A SHOT.

Also the pressure of the wedge **on** the plane

$$= Mg + (R_1 + R_2) \cos \alpha + 2T \sin \alpha$$

$$= \frac{M + m_1 + m_2}{M + (m_1 + m_2) \sin^2 \alpha} \left[ M + \frac{4 m_1 m_2 \sin^2 \alpha}{m_1 + m_2} \right] g.$$

Now $\dfrac{4 m_1 m_2 \sin^2 \alpha}{m_1 + m_2}$ is $< \dfrac{(m_1 + m_2)^2 \sin^2 \alpha}{m_1 + m_2}$ is $< (m_1 + m_2) \sin^2 \alpha.$

Hence this pressure is $\quad < (M + m_1 + m_2) g,$

or the pressure is less than it would be if the string were **fixed at some** point of its length so as to prevent motion and the wedge **kept at rest**.

**Ex. 4.** *Assuming that in a cannon the force on the ball depends only on the* space *occupied by the volume of the vapour of the gunpowder, shew that* **the ratio of the** *final velocity of the ball when the* **gun** *is free to recoil to its velocity* **when the gun is fixed is** $\sqrt{\dfrac{M}{M+m}}$, *where* **M, m** *are the masses of the cannon and ball respectively.*

When the gun is free to recoil let the velocities of the ball and gun be $U$, $V$ initially.

Then since the total momentum generated in the gun in one direction is equal to the momentum generated in the shot in the opposite direction, we have

$$mU = MV \dotfill (1).$$

Also since the change in the kinetic energy of a system is equal to the work done on it, we have

$$\tfrac{1}{2} m U^2 + \tfrac{1}{2} M V^2 = \text{work done by the explosion} \dotfill (2).$$

When the gun is fixed let $U_1$ be the initial velocity of the shot. Then

$$\tfrac{1}{2} m U_1^2 = \text{work done by the explosion in the second case} \dotfill (3).$$

Now since the pressure of the vapour of the powder depends *only* on the space it occupies, **and** since in each case the powder expands from the volume it occupied originally to the **volume that the** gas occupies in the ordinary **atmosphere**, the work done in **the two cases is the same**.

$$\therefore \ \tfrac{1}{2} m U_1^2 = \tfrac{1}{2} m U^2 + \tfrac{1}{2} M V^2,$$

and, substituting from (1) for $V$, we have $Mm U_1^2 = Mm U^2 + m^2 U^2.$

$$\therefore \ \frac{U^2}{U_1^2} = \frac{M}{M + m}.$$

In this example the mass of the products of the explosion may be either looked upon as negligible; or it may be assumed that they are expelled with the same velocity as the bullet and their mass included in that of the bullet.

**Ex. 5.** *If* $a$ *be the penetration of a shot of* $m$ *lbs. striking a fixed iron plate with velocity* $v$, *shew that a plate of* $M$ *lbs. and thickness* $b$, *free to move, will be perforated if* $b < \dfrac{M}{M+m} a$, *and that after passing through the plate the shot will retain the velocity* $\dfrac{m + \sqrt{M^2 - M(M+m)\dfrac{b}{a}}}{M+m} v$, *the resistance being supposed uniform.*

Let $F$ be the resistance. Then since the shot is reduced to rest in passing through a distance $a$ of the plate, we have

$$\tfrac{1}{2} mv^2 = F \cdot a \quad \dotfill (1).$$

Let $x$ be the thickness of the thinnest plate, free to move, which will just stop the bullet. When the distance $x$ has been traversed the bullet and plate are moving with a common velocity. Call this velocity $U$. Then since the total momentum is unchanged by the impact, we have

$$mv = (m + M) U \quad \dotfill (2).$$

Also since the loss of kinetic energy is equal to the work done we have

$$\tfrac{1}{2} mv^2 - \tfrac{1}{2}(m+M) U^2 = Fx = \frac{x}{a} \tfrac{1}{2} mv^2 \quad \dotfill (3).$$

From (2) and (3) we have, by eliminating $U$,

$$x = a \frac{M}{M+m},$$

so that the plate will be perforated if the thickness $b$ be $< \dfrac{M}{M+m} a$.

If the shot pass through the plate, let its velocity on emergence be $V$. The equations (2) and (3) now become

$$mv = mV + MU,$$

and $\quad \tfrac{1}{2} mv^2 - \tfrac{1}{2} mV^2 - \tfrac{1}{2} MU^2 = Fb = \tfrac{1}{2} m \dfrac{b}{a} v^2,$

∴ eliminating $U$ we have

$$Mmv^2 - MmV^2 - m^2(v - V)^2 = Mm \frac{b}{a} v^2.$$

$$\therefore V^2(M+m) - 2mvV = v^2 \left[ M - m - M \frac{b}{a} \right].$$

PILE-DRIVING. 121

Whence by solving

$$V = \frac{m + \sqrt{M^2 - M(M+m)\frac{b}{a}}}{M+m} v.$$

Ex. 6. *An inelastic pile of mass* m *lbs. is driven vertically* a *feet into the ground by* n *blows of a hammer of* M *lbs. falling vertically through* h *feet; shew that* $\frac{nM^2}{M+m}\frac{h}{a}$ *lbs. superposed on the pile in addition to* M *would drive it down very slowly supposing the resistance uniform.*

*If the pile be crushed* x *feet by each blow, where* x *is small, the mean* **pressure** *exerted by the hammer is equal to the weight of* $\frac{Mm}{M+m}\frac{h}{x}$ *pounds, and each blow lasts for* $\frac{x}{h}$ **of the time of falling of** *the* **hammer,** *neglecting forces not due to the impulse.*

Let $V_1$ be the velocity **of the pile and** hammer immediately after the blow. Since the total momentum **is** unaltered by the blow,

$$M\sqrt{2gh} = (M+m)V_1 \dots\dots\dots\dots\dots\dots(1).$$

Let $F$ be the resistance of the **ground.** Then the total **force on** the pile is $F - (M+m)g$, and this **force** brings the pile and hammer **to** rest in a space $\frac{a}{n}$.

$$\therefore \tfrac{1}{2}(M+m)V_1^2 = [F - (M+m)g] \cdot \frac{a}{n}.$$

$$\therefore F - (M+m)g = \frac{n}{a}(M+m)\frac{V_1^2}{2} = \frac{ngh}{a}\frac{M^2}{M+m}.$$

$\therefore$ the resultant upward pressure = weight of $\frac{nM^2}{M+m}\frac{h}{a}$ pounds so that this mass superimposed would just overcome the resistance and force the pile slowly downwards.

Again during the impact let $P$ denote the mean pressure between the hammer and the pile. Then since the change in the kinetic energy of the hammer is **equivalent to** the work done, we have

$$\tfrac{1}{2}M \cdot 2gh - \tfrac{1}{2}(M+m)V_1^2 = Px.$$

Substituting for $V_1$ from (1), we have $P = \frac{Mm}{M+m}\frac{h}{x}g$,

or mean pressure = weight of $\frac{Mm}{M+m}\frac{h}{x}$ pounds.

Also time of each blow

$$= \frac{\text{change in the momentum of the hammer}}{P}$$

$$= \frac{M\sqrt{2gh} - MV_1}{P} = \frac{M(M+m)\sqrt{2gh} - M^2\sqrt{2gh}}{Mm\frac{h}{x}g}$$

$$= \frac{x}{h}\sqrt{\frac{2h}{g}} = \frac{x}{h} \times \text{time of falling of the hammer.}$$

Ex. 7. *Two men, each of mass* M, *stand on two inelastic platforms, each of mass* m, *hanging over a small smooth pulley. One of the men leaping from the ground would be able to raise his centre of* inertia *through a distance* h; *shew that if he leap with the same energy from his platform his centre of inertia would rise a distance* $\frac{M+2m}{2(M+m)}$ h, *and that he would fall on the platform in its original position and the whole system be again at rest.*

Since the man can raise his centre of inertia through a height $h$, the work he does on his system by leaping must be $Mgh$.

Let $U$, $V$ be the initial velocities of the platform and the man.

Then since the momentum of the man is equal and opposite to that of the system consisting of the platforms and the other man,

$$\therefore\ MV = (M+2m)U \quad\ldots\ldots\ldots\ldots\ldots\ldots (1).$$

Also since he leaps with the same energy in the two cases we have

$$\tfrac{1}{2}MV^2 + \tfrac{1}{2}(M+2m)U^2 = Mgh\ldots\ldots\ldots\ldots\ldots(2).$$

Solving these equations we have

$$V^2 = \frac{M+2m}{M+m}gh.$$

Hence height through which his centre of inertia rises

$$= \frac{V^2}{2g} = \frac{M+2m}{2(M+m)}h.$$

Also time before he returns to his original position $= 2\dfrac{V}{g}\ldots\ldots\ldots(3)$.

After he has left the platform it starts with initial velocity $U$ downwards and moves with an acceleration $\dfrac{M}{M+2m}g$ in the opposite direction.

## MOTION OF AN ELASTIC RING ON A CONE.

∴ if $T$ be the time before the platform is in its original position again, we have

$$0 = UT - \tfrac{1}{2}\frac{M}{M+2m}gT^2.$$

$$\therefore T = \frac{2(M+2m)}{M}\frac{U}{g} = \frac{2V}{g}, \text{ by (1)}.$$

Hence the **man** and his platform are in their original positions again at the same instant, and since their momenta are then equal and opposite the whole **system** comes to rest.

Ex. 8. *A heavy elastic ring in the form of a circle is placed with its plane horizontal over a smooth cone whose axis is vertical; if the ring be held in contact with the cone at its natural length and allowed to fall, how far will it descend before* **first coming** *to rest?*

Since the **change** in the **kinetic energy** of a system is equal to the work done on it, and since the **kinetic** energy of the ring **both at the** beginning and end of the motion **is zero** and hence the total change zero, the total amount of the work done on the ring during the motion must be zero also.

Let $2\alpha$ be the **vertical angle of the cone**, $2\pi a$ the unstretched length of the ring, $\lambda$ its coefficient **of elasticity, and let** $2\pi x$ be the length of the ring when it first comes to rest.

The work done by the **weight of the ring during the motion** is $W(x-a)\cot\alpha$, where $W$ is its weight.

Also the work done by the tension is (Art. 91, **Ex. 2**),

$$-\tfrac{1}{2}(2\pi x - 2\pi a)\lambda\frac{x-a}{a}, \quad \text{or} \quad -\frac{\pi\lambda}{a}(x-a)^2.$$

Also **the work done** by the normal **pressure of the cone is zero**. (Art. 89.)

Hence we have $\quad W(x-a)\cot\alpha - \dfrac{\pi\lambda}{a}(x-a)^2 = 0$.

$$\therefore x-a = \frac{Wa\cot\alpha}{\pi\lambda}, \text{ and distance the string descends} = \frac{W}{\lambda}\frac{a}{\pi}\cot^2\alpha.$$

Ex. 9. *A goods* **train** *consists of a number of similar waggons* **and an** *engine whose mass is* **an** *integral number* ($\mu$) *times the mass of a waggon. They are coupled by chains of equal length, inelastic, and weightless. The train is initially* **at rest on** *a straight* **line** *of railway with all the vehicles in contact and the* **couplings slack.** *The engine* **then** *begins to move, the* *steam exerting a constant* tractive force, and each waggon starting with a jerk as *its coupling* tightens. *Shew* that, during the starting, the velocity of the moving part of the *train will* be greatest just before the nth impact, *where* $n = 2\mu^2 - 3\mu + 1$, *provided there are at least this number of impacts. All rotation of the wheels and friction may be neglected.*

Let $V_n$ be the velocity of the train just before the $n$th truck starts, $U_n$ its velocity just after the $n$th truck starts.

Then since the total momentum of the train is unaltered by the jerk which sets the $n$th truck in motion, we have

$$(\mu + n - 1)V_n = (\mu + n)U_n. \quad \ldots\ldots\ldots\ldots (1).$$

Also if $F$ be the force in poundals exerted by the engine and $m$ the mass of a truck, the acceleration when $n$ trucks are in motion is $\dfrac{F}{(n+\mu)m}$.

Hence since the train moves for a distance $a$ with this acceleration,

$$V_{n+1}^2 = U_n^2 + 2\dfrac{F}{(n+\mu)m}a \ldots\ldots\ldots\ldots (2),$$

where $a$ is the length of a coupling. Hence from (1) (2)

$$(\mu+n)^2 V_{n+1}^2 - (\mu+n-1)^2 V_n^2 = \dfrac{2Fa}{m}(\mu+n).$$

So $\quad (\mu+n-1)^2 V_n^2 - (\mu+n-2)^2 V_{n-1}^2 = \dfrac{2Fa}{m}(\mu+n-1).$

$$\ldots\ldots\ldots\ldots\ldots\ldots\ldots\ldots\ldots\ldots\ldots\ldots\ldots\ldots$$

$$(\mu+1)^2 V_2^2 - \mu^2 V_1^2 = \dfrac{2Fa}{m}(\mu+1).$$

But $V_1$ = velocity of the engine just before the first truck moves

$$= \sqrt{\dfrac{2F}{\mu m}a}.$$

∴ adding and substituting for $V_1$ we have

$$(\mu+n)^2 V_{n+1}^2 - \dfrac{2Fa}{m}\mu = \dfrac{2Fa}{m}\left[\dfrac{n(n+1)}{2} + n\mu\right].$$

$$\therefore (\mu+n)^2 V_{n+1}^2 = \dfrac{Fa}{m}(2\mu+n)(n+1).$$

So $\quad V_n^2 = \dfrac{Fa}{m}\dfrac{(2\mu+n-1)n}{(n+\mu-1)^2} = \dfrac{Fa}{m(\mu^2-\mu)}\left\{\mu^2 - \mu + \tfrac{1}{4} - \left[\dfrac{\mu^2-\mu}{n+\mu-1} - \tfrac{1}{2}\right]^2\right\}.$

Hence the greatest value of $V_n$ is when $\dfrac{\mu^2-\mu}{n+\mu-1} - \tfrac{1}{2} = 0$, or when $n = 2\mu^2 - 3\mu + 1$.

103. We shall now solve some questions relating to the motion of heavy strings. The best method for a student to obtain a clear idea of the motion is to consider the case of a light string on which are fixed a number of masses, equal to one another, and then to consider the limiting case when the distances between the particles become indefinitely small. In this manner we pass to the case of a uniform heavy string. If the string be not of uniform density, but have its density given according to some known law, then the masses should be so chosen that ultimately they may form a string of the required density. For example, if the heavy string be given to be of density which varies as its distance from one end, the masses should be chosen so as to be in the ratio of 1, 2, 3, ...; if it vary as the square of the distance from one end, the masses should be chosen in the ratios of $1^2$, $2^2$, $3^2$, ...; similarly any other case may be considered.

When a string is in motion there being no jerks (as in the case of a string stretched tight on a table and drawn over the edge by a portion which hangs vertically) the principle of the Conservation of Energy may be applied.

**Ex. 10.** *A heavy uniform string is coiled up on the edge of a horizontal table and a small portion hangs over the edge; if motion be allowed to ensue, shew that the end descends with uniform acceleration* $\tfrac{1}{3}g$, *and find the tension at the edge at any* **time.**

Consider the case of a light string laden with equal masses $m$ at equal intervals $a$.

Let $V_n$, $U_{n+1}$ be respectively the velocities of the system just before and after the $(n+1)$th particle comes over the edge.

Since the total momentum of the system is unaltered by the jerk necessary to set the $(n+1)$th particle in motion,

$$\therefore (n+1)U_{n+1} = nV_n \quad \text{...........................(1).}$$

Also between successive jerks the system moves freely with acceleration $g$,

$$\therefore V_n^2 = U_n^2 + 2ga \quad\text{............................(2)}.$$

Hence from (1) and (2)

$$(n+1)^2 U_{n+1}^2 - n^2 U_n^2 = 2ga \cdot n^2.$$

So

$$n^2 U_n^2 - (n-1)^2 U_{n-1}^2 = 2ga \cdot (n-1)^2,$$

$$\text{..........................................}$$

$$2^2 U_2^2 - 1^2 \cdot U_1^2 = 2ga \cdot 1^2.$$

Therefore by addition

$$(n+1)^2 U_{n+1}^2 - 1^2 U_1^2 = 2ga \cdot \frac{n(n+1)(2n+1)}{6}.$$

But $U_1$ = initial velocity of the first mass = 0.

$$\therefore U_{n+1}^2 = \frac{ga}{3} \frac{n(2n+1)}{n+1}.$$

Now let $na = x$, so that $U_{n+1}^2 = \frac{g}{3} \frac{x(2x+a)}{x+a}$.

If we make $a$ very small we have the case of a uniform heavy chain so that if $U_x$ denote the velocity when a distance $x$ of this heavy chain has passed over the table we have

$$U_x^2 = \frac{g}{3} 2x = \frac{2g}{3} x, \text{ (by neglecting } a\text{)}.$$

Hence the end of the chain is descending with a constant acceleration $\frac{g}{3}$.

Let $M$ be the mass of the string which is hanging vertically at any instant and $T$ the tension at the edge of the table.

Then $Mg - T$ = moving force on this portion = $M\frac{g}{3}$.

$$\therefore T = \frac{2Mg}{3} = \text{two thirds of the weight of the portion which hangs vertically.}$$

**Ex. 11.** *One end of a heavy uniform string hangs over a small pulley and the other end is coiled up on a table; if the part hanging over the pulley be initially of length* b *and the distance of the pulley from the table be* c *(< b) and motion ensue, find the velocity at any time and the tension of the string at the table.*

Consider the case of **a light** string with **masses** $m$ attached to it at equal intervals $a$; initially let $p$ of these be **hanging** on one side and $q$ on the other.

## MOTION OF HEAVY STRINGS.

Let $V_n$, $U_{n+1}$ be respectively the velocities of the system just before and after the $(n+1)$th mass passes over the pulley; also let us suppose the distance $a$ so chosen that as one mass passes over the pulley another is lifted from the table.

Then since the total momentum of the system is unaltered by the jerk necessary to set this particle in motion,

$$\therefore (n+p+1)\, U_{n+1} = (n+p)\, V_n \quad \ldots\ldots\ldots\ldots\ldots\ldots (1).$$

Now when $n$ particles are on one side of the pulley, and $p$ on the other the acceleration of the system $= \dfrac{n-p}{n+p} g$, and for a distance $a$ the system is moving with this acceleration.

$$\therefore V_n^2 = U_n^2 + 2\dfrac{n-p}{n+p} ga \quad \ldots\ldots\ldots\ldots\ldots\ldots (2).$$

From (1), (2) we have

$$(n+p+1)^2\, U_{n+1}^2 - (n+p)^2\, U_n^2 = 2ga\, (n^2 - p^2).$$

So
$$(n+p)^2\, U_n^2 - (n+p-1)^2\, U_{n-1}^2 = 2ga\{(n-1)^2 - p^2\},$$

$$\ldots\ldots\ldots\ldots\ldots\ldots\ldots\ldots\ldots\ldots\ldots\ldots\ldots\ldots\ldots\ldots$$

$$(q+p+1)^2\, U_{q+1}^2 - (q+p)^2\, U_q^2 = 2ga\,\{q^2 - p^2\}.$$

Also $U_q$, being the initial velocity of the system, is zero.

Hence, by addition,

$$(n+p+1)^2\, U_{n+1}^2 = 2ga\,[q^2 + (q+1)^2 + \ldots + n^2 - (n-q+1)\,p^2],$$

$$= 2ga\left[\dfrac{n(n+1)(2n+1)}{6} - \dfrac{(q-1)\,q\,(2q-1)}{6} - (n-q+1)\,p^2\right].$$

Now put $na = x$, $pa = c$, $qa = b$.

Then this equation is

$$(x+c+a)^2\, U_{n+1}^2 = 2g\left[\dfrac{x(x+a)(2x+a)}{6} - \dfrac{b(b-a)(2b-a)}{6} - c^2(x-b+a)\right].$$

**If we now** make $n$ indefinitely great and $a$ indefinitely small we obtain the case of the question in which we have a uniform heavy chain.

If $U_x$ be the velocity when a distance $x - b$ has passed over the pulley we obtain

$$(x+c)^2\, U_x^2 = 2g\left[\dfrac{x^3 - b^3}{3} - c^2(x-b)\right] = \dfrac{2g}{3}(x-b)\,[x^2 + xb + b^2 - 3c^2].$$

Let $T$ be the tension of the string at the table. Then the impulse of this tension in a small time $\tau$ is $T\tau$. Also in this time a portion of the string equal to $m\, U_x \tau$ is put into motion, where $m$ is the mass of the string per unit of its length; also its velocity is changed from 0 to $U_x$; hence the momentum generated is $m U_x^2 \tau$. But the impulse of a force is equivalent to the momentum generated, so that $T = m U_x^2$.

**104. Transmission of power by belts and shafts.** When a moving belt passes over a pulley, the work done by the belt on the pulley can be easily found. For let $T_1$, $T_2$ be the tensions of the two portions of the belt at the points where it meets and leaves the pulley respectively; also let $v$ be the velocity of any point of the belt in feet per second. Then the work done by the tension $T_2$ is $T_2 v$ and the work done against the tension $T_1$ is $T_1 v$. Hence the total work done on the pulley per second is

$$(T_2 - T_1) v.$$

As an example, let us find the horse-power transmitted to a shaft by a band which travels at the rate of 44 feet per second round a pulley which is firmly keyed to the shaft, when the tensions of the two portions of the band are 600 and 200 lbs. weight respectively.

Here $\qquad T_2 = 600\,g$; $T_1 = 200\,g$; $v = 44$.

∴ work done per second $= 400\,g \cdot 44 = 400 \times 44$ foot-pounds.

Hence the horse-power transmitted $= \dfrac{400 \times 44}{550} = 32$.

**105. When a couple of given moment is applied to a shaft which is turning with a given angular velocity the horse-power transmitted is known.**

Let $\omega$ be the angular velocity of the shaft, $G$ the moment of the couple applied to it, and $r$ the radius of the section of the shaft.

The couple $G$ may be represented by two forces $\dfrac{G}{2r}$ acting at the ends of a diameter. In the unit of time the point of application of each force moves through a distance $r\omega$. Hence the work done by each force is $\dfrac{G}{2r} \cdot r\omega$ or $\dfrac{G}{2}\omega$, and therefore the work done by the couple is $G \cdot \omega$.

Hence the work done in one second is $\dfrac{G\omega}{g}$ foot pounds and the horse-power transmitted is $G\omega \div 550\, g$.

For example, if a shaft be acted on by a couple of 10000 foot-pounds and is revolving 50 times per minute the work done on it per minute is

$$10000 \times 100\pi \text{ foot-pounds.}$$

Hence the horse-power transmitted is $\dfrac{10000 \times 100\pi}{33000}$ or $95\tfrac{5}{21}$.

## EXAMPLES. CHAPTER III.

1. In a system of pulleys (three moveable) in which all the strings are attached to the roof, the highest string after passing over a fixed pulley has a mass $m$ attached to it, and the lowest pulley a mass $M$. If $M$ descend shew that its acceleration is $(M - 8m)\,g/(M + 64m)$.

How is this expression altered if each pulley have a mass $\mu$?

2. In the system of pulleys where each string is attached to a bar which supports the weight, if there be two moveable pulleys, and the power be quadrupled, the weight will ascend with acceleration $\dfrac{3g}{29}$.

3. In a system of pulleys with $n$ strings at the lower block the upward acceleration due to a power $P$ is

$$(nP - W)\,g/(n^2 P + W).$$

If, when $W$ has an upward velocity $v$, the weight $P$ reach the ground there will presently be on the string a strain of impulse $nPWv/(n^2 P + W)$.

L. D.

4. Two weights, $P$ and $W$, equilibrate on a wheel and axle, the mass of which may be neglected. A weight $W$ is attached to $P$ and after the lapse of one second another weight $W$ is attached to the ascending weight $W$; shew that after the lapse of another second the velocity of the ascending weight $2W$ is $\dfrac{b(2a-b)}{a^2+ab+2b^2}g$, where $a$, $b$ are the radii of the wheel and axle respectively.

5. A fine string passes over a smooth fixed pulley, and carries at its ends two small smooth pulleys, each of mass one pound, and these in turn carry strings with masses 1 lb., 2 lbs. and 1 lb., 3 lbs. tied to their ends respectively. Shew that the acceleration of each of the moveable pulleys is $\dfrac{g}{23}$.

6. A string one end of which is fixed has slung on it a mass $m_1$ and then passes over a fixed pulley having a mass $m_2$ attached to it at the other end; find the accelerations of the system and the tension of the string.

7. A string carries at one end a mass $M$, and at the other a pulley of mass $M'$, and is placed over a fixed pulley; over the pulley $M'$ passes a string carrying at its ends masses $m$ and $m'$; shew that the acceleration of $M$
$$=\dfrac{4mm'+(M'-M)(m'+m)}{4mm'+(M'+M)(m'+m)}g.$$
Compare the spaces passed over by $M$ and $m$, in a given time, and shew that the ratio of the upper and lower tensions is
$$4mm'+M'(m+m') : 2mm'.$$

8. A string passes over a fixed pulley $B$, and has a mass $m$ at one end and a pulley, $C$, of mass $p$, at the other. A string is fastened to a fixed point $A$, vertically below $B$, passes over $C$, and has at its other end a mass $m'$; shew that the acceleration of the pulley is $(2m'-m+p)g/(m+4m'+p)$.

EXAMPLES ON CHAPTER III. 131

9. A smooth ring of mass $M$ is threaded upon a string whose ends are then placed over two smooth fixed pulleys with masses $m$, $m'$ tied at the other ends respectively, and the various portions of the strings hang vertically. Find the motion of the ring and the two weights, and shew that the ring will be at rest if $M = 4mm'/(m + m')$.

Explain how the question is changed if $M$ be fixed on the string.

10. One end of a string is fixed; it then passes under a moveable pulley to which a weight $W$ is attached, and then over a fixed pulley and a smaller weight $P$ is attached to its other end, all three sections of string being vertical; shew that, neglecting the mass of the pulley, the acceleration with which $W$ descends is $(W - 2P)g/(W - 4P)$.

11. Two pulleys whose weights are $W$ and $W'$ are connected by a string hanging over a smooth pulley; over the former is hung a string with weights $P$, $Q$ at its extremities and over the latter one with weights $R$, $S$. Shew that when the system is free to move the acceleration of the pulleys is $\dfrac{W - W' + 2H - 2H'}{W + W' + 2H + 2H'} g$, where $H$ is the harmonic mean of $P$, $Q$ and $H'$ that of $R$, $S$.

12. Over a pulley is placed a string at one end of which is a single weight; at the other end is another pulley over which is slung a string carrying masses $m$, $m'$ at its ends; find the magnitude of the single weight so that $m'$ may remain at rest, if initially so, and shew that the acceleration of the pulley is $\dfrac{1}{2} \dfrac{m' - m}{m} g$, if its weight be neglected.

13. The masses of three particles are $m$, $2m$ and $3m$; $m$ and $2m$ are joined by a string passing round a pulley and placed on a smooth table to the edge of which the strings are

9—2

perpendicular; $3m$ is joined by a string to the pulley, supposed weightless, and hangs over the edge; shew that the acceleration of the mass $3m$ is $\frac{90}{17}g$.

14. A weight $W_1$ is placed on a rough table and has tied to it a light string which passes over the edge of the table and supports a pulley of weight $W_2$; round this pulley hangs another light string which has attached to it two unequal weights $W_3$, $W_4$; find the acceleration of the different parts of the system and also the relation between the different weights and the coefficient of friction that $W_1$ may remain at rest.

15. A sailor of weight $W$ sits in a loop of rope attached to the lower block of weight $W'$ of that system of pulleys in which one rope passes round all the pulleys. The lower block is suspended by $n$ vertical chords. If $P'$ be the force the sailor must exert on the free end of the rope in order to maintain equilibrium, shew that if he exert a force $P$ he will begin to move with an acceleration $(n+1)(P-P')g/(W+W')$.

16. In any machine without friction and inertia a weight $P$ supports a weight both hanging by vertical strings; if these weights be replaced by $P'$ and $W'$, and if in the subsequent motion $P'$ and $W'$ move vertically, the centre of inertia of $P'$ and $W'$ will descend with acceleration

$$g\frac{[WP'-PW']^2}{(W'+P')(P^2W'+W^2P')}.$$

17. Two equal weights $P$ are attached to the ends of a string, six parts of which are vertical, and which passes over three fixed pulleys and under two moveable pulleys each of weight $P$; another string has its ends fastened to the centre of the two moveable pulleys and supports a pulley and weight of joint weight $2P$. When the whole is in equilibrium a downward blow is given to one of the two weights. Compare the velocities communicated to the various pulleys and weights.

EXAMPLES ON CHAPTER III.    133

18.  $2n$ small smooth rings are fixed at equal intervals in a horizontal circle and an endless string passed through them in order.  If the loops of the string between each consecutive pair of rings support pulleys of masses $P, Q, R \ldots$ respectively, the portions of string not in contact with the pulleys being vertical, shew that the pulley $P$ will descend with acceleration

$$g\left(\frac{1-n}{P}+\frac{1}{Q}+\frac{1}{R}+\ldots\ldots\right)\bigg/\left(\frac{1}{P}+\frac{1}{Q}+\frac{1}{R}+\ldots\ldots\right).$$

19.  A bullet of mass **one ounce is** fired from a rifle weighing 6 lbs. freely suspended and the initial velocity **of the bullet is found to be 1000 feet per second.  Find what the** initial velocity **of** the bullet would be if the rifle were rigidly held.

20.  Two bullets **of** the same size but **compounded of** metals whose densities **are** in the ratio **3 : 4 are** shot vertically upwards in **a vacuum at the same instant from** two similar **guns charged with** the same quantities of powder; one returns to the point **of** projection half a **minute** later than the other; find the velocities of projection.

21.  A shell of mass $M$ is moving with velocity $V$.  An **internal explosion** generates **an amount** $E$ of energy and **thereby breaks the** shell into masses whose ratio is $m_1$ to $m_2$.  The fragments continue to move **in the original line of motion of the** shell; shew that their velocities are

$$V+\sqrt{\frac{2m_2 E}{m_1 M}} \quad \text{and} \quad V-\sqrt{\frac{2m_1 E}{m_2 M}}.$$

22.  A bullet fired horizontally from a musket being supposed to pass perpendicularly through a target suspended **freely in the air; explain on the dynamical theory** of work **why** the greater the velocity **of** the bullet **the** less the displacement of the target.

## EXAMPLES ON CHAPTER III.

**23.** Shew that the resistance of wood is 204 lbs. weight to a nail of mass one ounce, supposing that a hammer of mass 1 lb. striking it with a velocity of 34 feet per second drives the nail one inch into a fixed block of wood.

If the block be free to move and be of mass 68 lbs., shew that the hammer will drive the nail only $\frac{64}{65}$ths of an inch.

**24.** Two equal saucers lie on a smooth horizontal table and a perfectly rough insect jumps from the centre of one to the centre of the other and then back to the centre of the first. Shew that the final velocities of the saucers are in the ratio $M + m : M$, where $M$ is the mass of each saucer and $m$ the mass of the insect.

**25.** Two buckets of given length are suspended by a fine inelastic string placed over a smooth pulley; at the centre of the base of one of the buckets a frog of given mass is sitting; at an instant of instantaneous rest of the buckets the frog leaps vertically upwards so as just to arrive at the level of the rim of the bucket; find the ratio of the absolute length of the frog's vertical ascent in space to the length of the bucket and shew that the time which elapses before the frog again arrives at the base of the bucket is independent of the frog's weight.

**26.** Two masses $m, m'$ are connected by an elastic string and are placed on a smooth horizontal table. Initially they are at rest and the string unstretched; a blow, whose impulse is $P$, is then given to the particle $m$ along the direction of the string; shew that when the string is next unstretched the velocity of $m'$ is $\dfrac{2P}{m + m'}$, and find the velocity of $m$ at the same instant.

**27.** Two particles $A, B$ of equal mass are connected by a rigid rod of negligible mass, and a third equal particle $C$ is tied to a point $P$ on the rod at distances $a$, $b$ from the two ends. $C$ is projected with velocity $u$ from $P$ perpendicular to the rod. Shew that just after the string becomes tight, $C$ will move with velocity $\dfrac{1}{2}\dfrac{a^2+b^2}{a^2+ab+b^2}u$, and find the initial velocities of $A$ and $B$.

28. Three equal particles are placed at the angular points of a triangle $ABC$ and $A$, $B$ are connected with $C$ by equal strings which are just taut. $C$ is projected perpendicularly towards $AB$. Find the loss of kinetic energy after $AB$ are jerked into motion.

29. A wedge of mass $M$ supports two masses $m$, $n$ one on each face and they are connected by a string passing over the summit and the whole is placed on a smooth table; shew that if during the motion the string break then the acceleration of $M$ will be instantaneously reversed in direction provided $\dfrac{m\sin\alpha}{n\sin\beta}$ lie between $\dfrac{\cos\beta}{\cos\alpha}$ and unity.

30. A shaft is transmitting a couple equal to 2000 foot-pounds and is making 150 revolutions per minute. What H.-P. does it transmit?

31. A shaft is driven by an engine through the intervention of a belt which travels at the rate of 30 feet per second, the tensions of the free portions of the belt being equal to the weights of 50 and 100 lbs. respectively. What is the horse-power transmitted?

32. A heavy wheel 5 feet in radius revolves on its axle which is horizontal and two inches in diameter making 10 turns per minute, its mass being supposed to be entirely collected at its rim. If left to itself the coefficient of friction being $\frac{1}{5}$, how many turns will it make before stopping? How

many turns will it make if it initially were making 20 revolutions per minute?

33. A uniform rectangular block $ABCD$ stands on its base $AD$ on a rough floor. It is pulled at $C$ by a horizontal force just great enough to begin to turn it round the corner $D$. If the same force continue to pull horizontally at $C$ till the block has turned through an angle $\theta$ and then cease to act, shew that the block will just have acquired sufficient momentum to cause it to turn over round $D$ provided that $\sin\theta = \tan\frac{a}{2}$ where $a$ is the angle $BDC$.

34. A heavy elastic ring in the form of a circle is placed with its plane horizontal over a smooth sphere whose axis is vertical; if the string be held in contact with the sphere at its natural length and allowed to fall, how far will it descend before first coming to rest? What is the condition that it should just slip over the sphere?

35. Two fixed points in the same horizontal line are connected by an elastic string whose natural length is the distance between the points and whose coefficient of elasticity $\frac{w}{\sqrt{3}}$; a particle of weight $w$ is fastened to the middle point of the string. If the particle be held so that the string is unstretched and be let go, shew that the angle $2\theta$ between the two strings when the particle has fallen to its greatest depth is given by the real root of the equation $4\sin^3\theta + 3\sin\theta = 1$.

36. Two strings each of length $a$ are fastened to a particle of mass $m$. The one end of one string is fastened to a fixed point $A$, and the other passes over a small smooth pulley $B$ in the same horizontal plane with $A$ and carries a particle of mass $m'$. The distance $AB$ is equal to $a$. If $m$ be

originally at rest and very near $B$, the string being tight, shew that just before $m'$ reaches $B$ the velocity of $m'$ is

$$\sqrt{\frac{m\sqrt{3}-2m'}{\frac{4}{3}m+m'}ga}.$$

What happens if $m$ be $<\dfrac{2}{\sqrt{3}}m'$?

37. Two equal masses, each equal to $P$, balance by means of a string connecting them which passes over two pulleys, $A$, $B$ in the same horizontal line. If a third mass $M$ be placed at $C$, the middle point of $AB$, it will descend until

$$\tan\frac{BAC}{2}=\frac{W}{2P}.$$

Shew also that if $a$ be the value of the angle $BAC$ at any time the downward velocity of the mass $M$ will be

$$\left\{2ag\,\frac{M\tan a-2P(\sec a-1)}{M+2P\sin^2 a}\right\}^{\frac{1}{2}}$$

where $AB=2a$.

38. A railway **train** consisting of **a number of carriages, is so** coupled **that there is a** distance **of, say, three feet between each** carriage and the **next. Explain** why the resistance which such a train experiences from a cross-wind should be greater than from a head-wind when the train is going at full speed.

If the transverse section **of the carriages be a square of 7** feet side, **the velocity of the** wind 45 miles per hour, the velocity of **the train** 60 miles per hour, and there be 10 spaces each **3 feet wide** between the carriages, shew **that** the resistance due to the wind is equal to the weight of more **than** a ton, the **mass of a cubic foot of** air being taken to be $1\frac{1}{4}$ ozs.

39. Two particles, each of mass $M$, are **tied one** at each end of an inelastic string of length $2a$, and a particle of mass

$m$ is tied at its middle point. The system is laid on a smooth horizontal plane with the string at its full length in a straight line. A blow, of impulse $F$, perpendicular to the string is given to the middle particle. Shew by means of the equations of Energy and Momentum that, when the two portions of the string each make an angle $\theta$ with their initial positions, the angular velocity of either is $\dfrac{F}{a\sqrt{m(2M\sin^2\theta + m)}}$.

40. A wedge, whose section is a right-angled triangle, rests on a rough table with the hypotenuse inclined to the table at an angle $i$. A heavy smooth chain of length $a$ is fastened to the wedge near the right angle, and passing over a smooth pulley hangs over the edge of the table, so that the wedge is just on the point of motion in the direction of the pulley when a length $b$ of the chain hangs over the table. A heavy particle is then placed on the inclined face of the wedge and slides down it. Shew that the particle will separate from the wedge when the end of the chain has descended through a distance $\left(\dfrac{b}{\mu} + a\right)\cot i$, where $\mu$ is the coefficient of friction between the wedge and the table, the motion being supposed to take place in a vertical plane.

41. An umbrella, whose surface is smooth and spherical, is held in rain which is falling vertically with velocity $v$. The umbrella itself is being drawn vertically downwards with velocity $V$. Prove that, if $V$ be less than $v$, the average pressure per unit area of the rain falling on the umbrella at a point whose angular distance from the highest point is $\theta$, is $p\,\dfrac{(v-V)^2\cos^2\theta}{v^2}$, where $p$ is the average pressure per unit area of the rain falling on a fixed horizontal plane.

42. A large number of equal particles are fastened at unequal intervals to a fine string, and then collected into a

heap at the edge of a smooth horizontal table with the extreme one just hanging over the edge. The intervals are such that the times between successive particles being carried over the edge are equal; shew that if $c_n$ be the interval between the $n$th and $(n+1)$th, and $v_n$ be the velocity just after the $(n+1)$th particle is carried over, then $\dfrac{c_n}{c_1} = \dfrac{v_n}{v_1} = n$.

43. $n$ equal masses are placed in contact in a line on a smooth horizontal table, each being connected with the next by an inelastic string of length $a$, and another equal mass is attached to the foremost of these masses by a string which passes over a pulley at the edge of the table. Shew that, of the kinetic energy generated until the last mass is set in motion the fraction $\dfrac{n}{2(n+1)}$ is lost, supposing none of the masses to leave the table until the last is set in motion.

44. A fine inelastic thread is loaded with $n$ equal particles at equal distances $c$ from one another; the thread is stretched and placed on a smooth horizontal table perpendicular to its edge over which one particle just hangs; shew that the velocity of the system when the $r$th particle is just leaving the plane is $\sqrt{gc\dfrac{r(r-1)}{n}}$.

Hence shew, that if a heavy string of length $a$ be similarly placed on a smooth table, the velocity with which it leaves the table is $\sqrt{ag}$.

45. $n$ particles are attached to a string at equal distances; it is placed so that $r$ of the particles are on an inclined plane and $n - r$ hang vertically; the system is set in motion by drawing one of the particles just over; shew that the velocity of the system when the last particle just leaves the plane is $\sqrt{(r-1)gc}$, where $c$ is the length of the string between consecutive particles.

46. A number of equal particles are fastened at equal distances $a$ on a string and placed in contact in a vertical line; shew that if the lowest be then allowed to drop, the velocity with which the $n$th begins to move is

$$\sqrt{ag\frac{(n-1)(2n-1)}{3n}}.$$

47. An indefinite quantity of a uniform chain is coiled in a heap on the floor of a room, and escapes into the room below through a hole in the floor; shew that the velocity of escape can never exceed $\sqrt{ga}$, where $a$ is the height of the hole above the floor of the room below.

48. A piece of uniform chain hangs vertically from its upper end, the lower end just touching a smooth inclined plane initially; if it be let go shew that at any instant during the motion $\frac{2}{3}$rds of the pressure on the plane is due to the impacts.

49. A fine uniform string of length $2a$ is in equilibrium passing over a small smooth pulley and is just displaced; shew that the velocity of the string when just leaving the pulley is $\sqrt{ag}$.

50. A chain of length $a$ is coiled up on a ledge at the top of a rough inclined plane, and one end is allowed to slide down. If the inclination of the plane is double the angle of friction $\lambda$, the chain will be moving freely at the end of time

$$\sqrt{\frac{6a\cot\lambda}{g}}.$$

51. A fine uniform chain is collected into a heap on a horizontal table; to one end is attached a fine string which, passing over a small smooth pulley vertically above the chain, carries a weight equal to that of a length $a$ of the chain;

shew that the length of the chain raised before the weight comes to rest is $a\sqrt{3}$, and find the subsequent motion.

52. A large number of equal particles are attached at equal intervals, $a$, to a fine string which passes through a short fine semi-circular tube, and initially $2r$ particles are on one side, the highest being at the tube, and $r$ particles on the other, the lowest being in contact with a horizontal table where the other particles are heaped up; shew that the velocity just before the $n$th additional particle is set in motion is

$$\sqrt{\frac{nag}{3}\left\{2+\frac{n-1}{(n+3r-1)^2}\right\}},$$

and deduce the corresponding result for a uniform chain hanging over a pulley.

53. A uniform heavy chain is vertical, both ends being fixed. The upper end $A$ is then released and the chain falls past the lower end $B$ without striking it. At the instant that $A$ is passing $B$, the latter is released. Shew that the chain will become straight again in $\tfrac{3}{4}$ of the time in which $A$ fell to $B$.

54. A uniform heavy chain is coiled up in a man's hand and one end tied to a fixed point. If the hand be suddenly removed, shew that the strain on the fixed point at any time during the motion is equal to three times the weight of the portion of chain hanging vertically at that time.

55. A length $l$ of uniform chain is lying in a heap on a horizontal plane, and a man takes hold of one end and continues to raise that end in a vertical line with uniform velocity $v$. Shew that the force exerted by the man is greater than the weight of the vertical portion at any time in the ratio $h+x:x$, where $x$ is the vertical portion and $h$ twice the height to which the velocity $v$ is due.

56. A uniform flexible chain of indefinite length, the mass of the unit of length of which is $m$, lies coiled on the ground, whilst another portion of the same chain forms a coil on a platform at a height $h$ above the ground, the intermediate portion passing round the barrel of a windlass placed above the second coil. An engine which can do $H$ units of work per unit of time is employed to wind up the chain from the ground and to let it fall into the upper coil. Shew that the velocity of the chain can never exceed the value of $v$ given by the equation $mghv + \tfrac{1}{2}mv^3 = H$.

57. A chain, whose density varies as the distance from one end, is coiled up close to the edge of a smooth table and the end allowed to hang over. Shew that the motion is uniformly accelerated, and that the tension at the edge of the table varies as the 4th power of the time that has elapsed since the commencement of the motion.

# CHAPTER IV.

### UNITS AND DIMENSIONS.

**106.** When we wish to state the magnitude of any concrete quantity we express it in terms of some unit of the same kind as itself, and we have to state, (1), what is the unit we are employing, and, (2), what is the ratio of the quantity we are considering to that unit. This latter ratio is called the *measure* of the quantity in terms of the unit. Thus if we wish to express the height of a man we may say that it is six feet. Here a foot is the unit and six is the measure. We might as well have said that he is 2 yards or 72 inches high.

The measure will vary according to the unit we employ. The measure of a certain quantity multiplied into the unit employed is always the same (e.g. 2 yards = 6 feet = 72 inches).

Hence if $k$, $k'$ be the measures of a physical quantity when the units used are denoted by $[K]$, $[K']$, we have

$$k[K] = k'[K']$$

and hence
$$[K]:[K'] :: \frac{1}{k} : \frac{1}{k'},$$

so that, by the definition of variation, we have $[K] \propto \frac{1}{k}$ or the unit in which any quantity is measured varies inversely as the measure and conversely.

**107.** A straight line possesses length only and no breadth or thickness, and hence is said to be of one dimension in length.

An area possesses both length and breadth, but no thickness, and is said to be of two dimensions in length. The unit of area usually employed is that whose length and breadth are respectively equal to the unit of length. Hence if we have two different units of length in the ratio $\lambda : 1$, the two corresponding units of area are in the ratio $\lambda^2 : 1$, so that if $[A]$ denote the unit of area and $[L]$ the unit of length, then

$$[A] \propto [L]^2.$$

A volume possesses length, breadth, and thickness, and is said to be of three dimensions in length. The unit is that volume whose length, breadth, and thickness are each equal to the unit of length. As in the case of areas it follows that if $[V]$ denote the unit of volume then

$$[V] \propto [L]^3.$$

Since the units of area and volume depend on that of length they are said to be **derived units**, whilst the unit of length is called a fundamental unit.

Another fundamental unit is the unit of time usually denoted by $[T]$. A period of time is of one dimension in time.

The third fundamental unit is the unit of mass denoted by $[M]$. Any mass is said to be of one dimension in mass.

These are the three fundamental units; all other units depend on these three, and are therefore derived units.

## UNIT OF VELOCITY.

**108. Theorem.** *To shew that the unit of velocity varies directly as the unit of length, and inversely as the unit of time.*

In one system let the units of length, time, and velocity be denoted by $[L], [T]$, and $[V]$, and in a second system by $[L'], [T'], [V']$; also let

$$[L'] = m[L], \text{ and } [T'] = n[T].$$

Then a body is said to be moving with the original unit of velocity

  when it describes a length $[L]$ in time $[T]$;

$\therefore$ with velocity $m[V]$

  when it describes a length $m[L]$ in time $[T]$;

$\therefore$ with velocity $\dfrac{m}{n}[V]$

  when it describes a length $m[L]$ in time $n[T]$;

$\therefore$ with velocity $\dfrac{m}{n}[V]$

  when it describes a length $[L']$ in time $[T']$.

But it is moving with velocity $[V']$ when it describes a length $[L']$ in time $[T']$.

$$\therefore [V'] = \frac{m}{n}[V].$$

$$\therefore [V'] : [V] :: \frac{[L']}{[L]} : \frac{[T']}{[T]}$$

$$:: \frac{[L']}{[T']} : \frac{[L]}{[T]};$$

$\therefore$ by the definition of variation, $[V] \propto \dfrac{[L]}{[T]}$.

L. D.

**109. Theorem.** *To shew that the unit of acceleration varies directly* **as the unit of length, and** *inversely as the* **square of the unit of time.**

Take the units of length and time as before, and let $[F]$, $[F']$ denote the corresponding units of acceleration.

Then **a body is said** to be moving
with the original unit of acceleration
  when a vel. of $[L]$ per $[T]$ is added on per $[T]$,
∴ with acceleration $m[F]$
  when a vel. of $m[L]$ per $[T]$ is added on per $[T]$,
∴ with acceleration $\dfrac{m}{n}[F]$
  when a vel. of $m[L]$ per $n[T]$ is added on per $[T]$,
∴ with acceleration $\dfrac{m}{n^2}[F]$
  when a vel. of $m[L]$ per $n[T]$ is added on per $n[T]$,
∴ with acceleration $\dfrac{m}{n^3}[F]$
  when a vel. of $[L']$ per $[T']$ is added on per $[T']$.
But now the body is moving with the new unit of acceleration $[F']$;

$$\therefore [F'] = \frac{m}{n^3}[F].$$

$$\therefore [F'] : [F] :: m : n^2$$

$$:: \frac{[L']}{[L]} \cdot \frac{[T]^2}{[T']^2}$$

$$:: \frac{[L']}{[T']^2} \cdot \frac{[L]}{[T]^2}.$$

$$\therefore [F] \propto [L][T]^{-2}.$$

## EXAMPLES.

*Ex.* 1. *Find the measure in the centimetre-minute system of the acceleration due to gravity, taking its measure in the foot-second system to be* 32·2, *and assuming a metre to be* 39·37 *inches.*

In a falling body a velocity of     32·2     ft. per sec. is added on per sec.,

∴ ..............................     $60 \times 32\cdot2$     ft. ... min..................per sec.,

∴ ..............................     $60^2 \times 32\cdot2$     ft. ... min. ...... ...... per min.,

..............................     $\dfrac{60^2 \times 32\cdot2 \times 12}{\cdot 3937}$     cms. per min. ......... per min.,

∴ required measure     $= \dfrac{3600 \times 12 \times 32\cdot2}{\cdot 3937} = 3533249.$

This may be more concisely put as follows:

Let $x$ be the new measure;

$$\therefore x[F'] = 32\cdot2\,[F].$$

$$x = 32\cdot2 \times \frac{[L][T]^{-2}}{[L'][T']^{-2}} = 32\cdot2 \times \frac{[L]}{[L']} \times \left[\frac{T'}{T}\right]^2$$

$$= 32\cdot2 \times \frac{12}{\cdot 3937} \times 60^2, \text{ as before.}$$

*Ex.* 2. *Convert a horse-power into* C.G.S. *units, assuming as a rough approximation that* 1000 *grammes* = 2¼ *lbs. and* 1 *centimetre* = ·4 *inch.*

We have 1 lb. = $\frac{4}{11}$ . $10^3$ grammes, and 1 ft. = 30 cms.

Also weight of 1 gramme = 981 dynes roughly.

Now the agent is dragging

    the weight of     33000     lbs.     through 1 foot in 1 min.,

∴ ...............     550     lbs.     ......... 1 foot ... 1 sec.,

∴ ...............     $30 \times 550$     lbs.     ......... 1 cm. ... 1 sec.,

∴ ...............     $\dfrac{30 \cdot 550 \cdot 5 \cdot 10^3}{11}$     grms. ......... 1 cm. ... 1 sec.,

∴ the agent drags a force equal to $\dfrac{30 \cdot 550 \cdot 5 \cdot 10^3}{11} \times 981$ dynes through 1 cm. per sec. and is therefore doing about $7\cdot4 \times 10^9$ C.G.S. units of work per second.

Ex. 3. If the unit of length be one mile, and the unit of time one minute, find the units of velocity and acceleration.

*Ans.* 88 feet per second; $\frac{22}{15}$ foot-second units.

**Ex. 4.** If the unit of velocity be a velocity of 30 miles per hour, and the unit of time be one minute, find the units of length and acceleration.

*Ans.* 880 yards; $\frac{11}{15}$ foot-second units.

**Ex. 5.** If the unit of mass be 1 cwt., the unit of force the weight of one ton, and the unit of length one mile, shew that the unit of time is $\frac{1}{2}\sqrt{33}$ seconds.

**110. Dimensions. Def.** *When we say that the dimensions of a physical quantity are* $\alpha$, $\beta$, $\gamma$ *in length, time, and mass respectively, we mean that the unit in terms of which the quantity is measured varies as*

$$[L]^\alpha [T]^\beta [M]^\gamma.$$

Thus the results of Arts. 108, 109 are expressed by saying that the dimensions of the unit of velocity are 1 in length and −1 in time; while those of the unit of acceleration are 1 in length and −2 in time.

The cases in Arts. 108, 109 have been fully written out, but the results may be obtained more simply as in the following article.

**111. (1) Velocity.** Let $v$ denote the numerical measure of the velocity of a point which undergoes a displacement whose numerical measure is $s$, in a time whose numerical measure is $t$, so that

$$s = vt.$$

If $[L]$ $[T]$ $[V]$ denote the units of length, time, and velocity respectively, we have

$$s \propto \frac{1}{[L]}; \quad t \propto \frac{1}{[T]}; \quad v \propto \frac{1}{[V]},$$

$$\therefore \frac{1}{[L]} \propto \frac{1}{[V][T]},$$

or

$$[V] \propto [L][T]^{-1}.$$

(2) **Acceleration.** Let $v$ denote the velocity acquired by a particle moving with acceleration $f$ for time $t$ so that
$$v = ft.$$
If $[F]$ denote the unit of acceleration we have
$$f \propto \frac{1}{[F]}.$$
$$\therefore \frac{1}{[V]} \propto \frac{1}{[F]}\frac{1}{[T]}.$$
$$\therefore [F] \propto [V][T]^{-1} \propto [L][T]^{-2}.$$

(3) **Density.** Let $d$ be the density of a body whose mass is $m$ and volume $u$ so that
$$m = du.$$
If $[D]$, $[U]$ denote the units of density and volume we have
$$d \propto \frac{1}{[D]}; \quad u \propto \frac{1}{[U]},$$
$$\therefore \frac{1}{[M]} \propto \frac{1}{[D][U]}.$$
$$\therefore [D] \propto [M][U]^{-1} \propto [M][L]^{-3}.$$

If the body be very thin so that it may be considered as a surface only, we see similarly that the unit of surface density
$$\propto [M][L]^{-2}.$$
So if the body be such that its breadth and thickness may be neglected, (so that it is a material line only) we have
$$\text{unit of linear density} \propto [M][L]^{-1}.$$

(4) **Force.** If $p$ be the force that would produce acceleration $f$ in mass $m$ we have
$$p = mf.$$
∴ if $[P]$ denote the unit of force we have
$$[P] \propto [M][F] \propto [M][L][T]^{-2}.$$

(5) **Momentum.** If $k$ be the momentum of a mass $m$ moving with velocity $v$ we have
$$k = mv.$$
∴ if $[K]$ denote the unit of momentum,
$$[K] \propto [M][V] \propto [M][L][T]^{-1}.$$

(6) **Impulse.** If $i$ be the impulse of a force $p$ acting for time $t$ we have
$$i = pt.$$
∴ if $[I]$ denote the unit of impulse,
$$[I] \propto [P][T] \propto [M][L][T]^{-1},$$
so that an impulse is of the same dimensions as a momentum.

(7) **Kinetic Energy.** If $e$ be the kinetic energy of a mass $m$ moving with velocity $v$ we have
$$e = \tfrac{1}{2}mv^2.$$
∴ if $[E]$ denote the unit of kinetic energy,
$$[E] \propto [M][V]^2 \propto [M][L]^2[T]^{-2}.$$

(8) **Work.** If $w$ be the work done when a force $p$ moves its point of application through a distance $s$ then
$$w = ps.$$
∴ if $[W]$ denote the unit of work,
$$[W] \propto [P][L] \propto [M][L]^2[T]^{-2}.$$
Hence work and kinetic energy are of the same dimensions.

(9) **Power** or Rate of work. If $h$ be the power at which work $w$ is done in time $t$, then

$$h = \frac{w}{t} = wt^{-1}.$$

∴ if $[H]$ denote the unit of power,

$$[H] \varpropto [W][T]^{-1} \varpropto [M][L]^2[T]^{-3}.$$

### EXAMPLES.

**1.** *If the acceleration of a falling body be taken as the unit of acceleration, and the velocity generated in a falling body in one minute as the unit of velocity, find the unit of length.*

In ft.-sec. units the velocity generated in one minute $= 60 \times 32$.

Let $[L], [T]$ denote a **foot and second** respectively, and $[L'], [T']$ the new units of length and **time**.

Then since the dimensions of an acceleration are $[L][T]^{-2}$, we have

$$1 \cdot [L'][T']^{-2} = 32[L][T]^{-2} \quad\quad\quad (1).$$

Also, since the dimensions of a velocity are $[L][T]^{-1}$, we have

$$1 \cdot [L'][T']^{-1} = 60 \cdot 32[L][T]^{-1} \quad\quad\quad (2),$$

dividing the **square of equation** (2) by (1) we have

$$[L'] = \frac{60^2 \cdot 32^2}{32}[L].$$

∴ the new unit of length $= 32 \times 60^2$ feet.

**2.** *The kinetic energy of a body expressed in the foot-pound-second system is* 1000; *find its value in the metre-gramme-minute system having given* 1 *foot* $= 30{\cdot}5$ *cms.*; 1 *lb.* $= 450$ *grammes approximately.*

Let $x$ be the measure in the new system, so that

$$x[E'] = 1000[E],$$

or $\quad\quad x[M'][L']^2[T']^{-2} = 1000[M][L]^2[T]^{-2}.$

But $\quad\quad [M] = 450[M'],\ [L] = {\cdot}305[L'],\ [T] = \tfrac{1}{60}[T']$

$$\therefore\ x = 1000 \times 450 \times ({\cdot}305)^2 \times 60^2$$
$$= 150{,}700{,}500.$$

## EXAMPLES.

**3.** *If the kinetic energy of a train of mass* 100 *tons and moving at the rate of* 45 *miles per hour be represented by* 11, *while the impulse of the force required to bring it to rest is denoted by* 5, *and* 40 *horse power by* 15, *find the units of time, length, and mass, and shew that the acceleration due to gravity is denoted by* 2016, *assuming its measure in foot-second units to be* 32.

In foot-second units the kinetic energy of the train $= \frac{1}{2} \cdot 100 \cdot 2240 \cdot 66^2$.
The impulse of the force required to bring to rest $= 100 \cdot 2240 \cdot 66$.
Also 40 H.-P. $= 40 \times 550 \times 32$ units of work per sec.
Hence since the dimensions of energy are $[M][L]^2[T]^{-2}$,

$$11[M'][L']^2[T']^{-2} = \tfrac{1}{2} \cdot 10^3 \cdot 224 \cdot 66^2 [M][L]^2[T]^{-2}\ldots\ldots\ldots(1).$$

Also since the dimensions of an impulse are $[M][L][T]^{-1}$,

$$\therefore 5[M'][L'][T']^{-1} = 10^3 \cdot 224 \cdot 66 [M][L][T]^{-1}\ldots\ldots\ldots\ldots(2).$$

Similarly considering the dimensions of a power,

$$15[M'][L']^2[T']^{-3} = 10^2 \cdot 55 \cdot 128 [M][L]^2[T]^{-3}\ldots\ldots\ldots\ldots(3).$$

Dividing (1) by (3), $\tfrac{11}{15}[T'] = \tfrac{1}{2} \dfrac{10^3 \cdot 224 \cdot 66^2}{10^2 \cdot 55 \cdot 128}[T]$.

$$\therefore [T'] = 21 \cdot 45 \text{ seconds} = 15\tfrac{3}{4} \text{ minutes}.$$

Also dividing (1) by (2), $\tfrac{11}{5}[L'][T']^{-1} = 33 \cdot [L][T]^{-1}$.

$$\therefore [L'] = [L] \cdot 15 \times 21 \cdot 45 = 105 \times 135 \text{ feet} = 2\tfrac{111}{176} \text{ miles}.$$

Also substituting in (2),

$$[M'] = \frac{10^3 \cdot 224 \cdot 66}{5} \cdot \frac{1}{105 \cdot 135} \cdot 21 \cdot 45 [M] = 88 \text{ tons}.$$

Hence the acceleration due to gravity

$$= 32 \frac{[L][T]^{-2}}{[L'][T']^{-2}} = 32 \frac{1}{105 \cdot 135} 21^2 \cdot 45^2 = 2016 \text{ new units}.$$

**4.** *If the unit of time be one minute, the unit of mass* 500 *pounds, and the unit of density that of water, find the unit of length, assuming that the mass of a cubic foot of water is a thousand ounces; shew also that the unit of force is about* $\tfrac{8}{15}$ *of a poundal.*

The new unit of density is $\tfrac{1000}{16}$ lbs. per cubic ft.

$$\therefore [D'] = \tfrac{1000}{16}[D], \text{ or } [M'][L']^{-3} = \tfrac{1000}{16}[M][L]^{-3}.$$

But $\qquad\qquad\qquad [M'] = 500[M].$

$$\therefore [L'] = 2[L] = 2 \text{ feet}.$$

Also $\quad\dfrac{[P']}{[P]}=\dfrac{[M']}{[M]}\dfrac{[L']}{[L]}\dfrac{[T']^{-2}}{[T]^{-2}}=500\cdot 2\cdot\dfrac{1}{60^2}.$

$\therefore [P']=\tfrac{5}{18}[P]=\tfrac{5}{18}$ of a poundal.

5. *Given that the unit of power is a million ergs per minute, that the unit of force is a thousand dynes, and the unit of time one-tenth of a second, what are the units of* **mass** *and length?*

Let $[M]$, $[L]$, $[T]$ denote a gramme, a centimetre and a sec., and $[M']$, $[L']$, $[T']$ the new units.

Since a million ergs per minute $=\dfrac{10^6}{60}$ ergs per second

$\qquad=\dfrac{10^5}{6}$ original units of power,

$\therefore 1\cdot[M'][L']^2[T']^{-3}=\dfrac{10^5}{6}[M][L]^2[T]^{-3}$ .............. (1).

Similarly since the new unit force $=10^3$. old unit force,

$\therefore 1\cdot[M'][L'][T']^{-2}=10^3[M][L][T]^{-2}$.................(2).

Also $\qquad\qquad [T']=\tfrac{1}{10}[T]$..................................(3).

Dividing (1) by (2) we have $\quad [L'][T']^{-1}=\dfrac{10^2}{6}[L][T]^{-1}$,

$\therefore$ by (3), $\qquad [L']=\dfrac{10^2}{6}[L]\cdot\tfrac{1}{10}=\tfrac{5}{3}$ centimetre.

Also substituting in (2),

$\qquad [M']\cdot\tfrac{5}{3}[L]\cdot 10^2[T]^{-2}=10^3[M][L][T]^{-2}$.

$\therefore [M']=6[M]=6$ grammes.

$\therefore$ the **required units** are 6 grammes and $\tfrac{5}{3}$ cm. respectively.

6. *A certain physical quantity is represented algebraically by a single term. If the unit of length be doubled the measure is $\tfrac{1}{4}$ of its former value, whilst if the units of length and time be both doubled the measure is also doubled. If in the former case the unit of mass be defined as the mass of unit volume of some substance, the measure is $\tfrac{1}{32}$nd of its former value. What kind of quantity is it?*

Let $a$, $\beta$, $\gamma$ be the dimensions of the quantity in mass, length, and time; and let $y$ be its measure when expressed in terms of the original unit $[Y]$ and let $[Y_1]$, $[Y_2]$, $[Y_3]$ be the units in the three following cases so that

$$y[Y]=\dfrac{y}{4}[Y_1]=2y[Y_2]=\dfrac{y}{32}[Y_3] \quad\ldots\ldots\ldots\ldots\ldots (1).$$

Let $[M']$ be the unit of mass in the last case. Then

$$[Y]:[M]^\alpha[L]^\beta[T]^\gamma :: [Y_1]:[M]^\alpha[2L]^\beta[T]^\gamma$$
$$:: [Y_2]:[M]^\alpha[2L]^\beta[2T]^\gamma :: [Y_3]:[M']^\alpha[2L]^\beta[T]^\gamma\ldots(2).$$

Now in the last case the unit of mass is defined as the mass of unit volume; and since the unit length has been altered in the ratio $1:2$, the unit volume has been altered in the ratio $1:8$;

$$\therefore [M']=8[M]\ldots\ldots\ldots\ldots\ldots\ldots\ldots\ldots(3).$$

Hence substituting from (2) in (1) and using (3), we have

$$1 = \tfrac{1}{4} \cdot 2^\beta = 2 \cdot 2^\beta \cdot 2^\gamma = \tfrac{1}{32} \cdot 8^\alpha \cdot 2^\beta,$$

$\therefore$ solving $\qquad \beta = 2; \quad \gamma = -3; \quad \alpha = 1.$

$$\therefore [Y] \propto [M][L]^2[T]^{-3},$$

so that the physical quantity is a power or rate of doing work.

7. Taking as a rough approximation 1 foot $=30{\cdot}5$ cms.; 1 lb. $=453$ grammes; and the acceleration of a falling body $=32$ ft.-sec. units, shew that

   (i) 1 Poundal $=13816$ Dynes,

   (ii) 1 Foot-Poundal $=421403$ Ergs,

   (iii) 1 Erg $=7{\cdot}416 \times 10^{-8}$ Foot-Pounds,

   (iv) 1 Horse-Power $=7{\cdot}416 \times 10^9$ Ergs per sec.,

   (v) 1 Horse-Power $=7{\cdot}6 \times 10^9$ gm.-cms. per sec.,

   (vi) 1 Foot-Pound $=13816$ gm.-cms.

8. **If the unit of** distance be one mile and the unit of time 4 **seconds** find the **units** of velocity and acceleration.

*Ans.* 1320 ft.-sec. units; 330 ft.-sec. **units.**

9. If the unit of acceleration be that of a freely falling body, and the **unit of time be 5** seconds, shew that the **unit** of velocity is a velocity of 160 ft. **per sec.**

10. What must be the unit of space, if the acceleration due to gravity be represented by 14, and the unit of time be five seconds?

*Ans.* $57\tfrac{1}{5}$ feet.

11. The acceleration **produced by gravity** being 32 in ft.-sec. units, find its measure when the **units are** $\tfrac{1}{10000}$ of an hour and a centimetre; given 1 centimetre $= {\cdot}0328$ ft.

*Ans.* $126\tfrac{8}{11}$.

EXAMPLES.   155

12. If the **area** of a ten acre field be represented by 100, and the acceleration of a heavy falling particle by 58¾, find the unit of time.

  *Ans.* 11 seconds.

13. If the unit of velocity be a velocity of 3 miles per hour, and the unit of time one minute, find the unit of length.

  *Ans.* 88 yards.

14. If the acceleration of a falling body be the unit of acceleration and the velocity acquired by it in 5 seconds be the unit of velocity, shew that the units of length and time are 800 feet and 5 seconds respectively.

15. If a hundredweight be the **unit of mass**, a minute the unit of time, and the unit of force the weight of a pound, find the unit of length.

  *Ans.* 342²⁄₇ yards.

16. If the units of velocity, length, and force be each doubled the units of time and mass will be unaltered, and that of energy increased in the ratio 1 : 4.

17. If the unit of work be that done in lifting one hundredweight through three yards, and the unit of momentum that of a mass of one pound which has fallen vertically 4 feet under gravity, and the unit of acceleration three times that produced by gravity, find the units of time, length, and mass.

  *Ans.* 21 secs.; 14112 yards; $\frac{1}{135}$ lb.

18. If 5½ yards be the unit of length, a velocity of one yard per second the unit of velocity, and 6 poundals the unit of force, what is the unit of mass?

  *Ans.* 11 lbs.

19. If the unit of time be one hour and the units of mass and force be the mass of one hundredweight, and the weight of a pound respectively, find the units of work and momentum in absolute units.

  *Ans.* The unit of work $= \frac{4}{7} \times 120^4$ absolute units of work; the unit of momentum $= 8 \times 120^2$ absolute units.

20. If the unit of force be the weight of 8 pounds, whilst a cubic foot of a substance of unit density is of mass 81 lbs., and the unit of time is one second, shew that the unit of length is 1 foot 4 inches.

21. If the units of force, work, and time be taken as the fundamental units, what are the dimensions of the unit of length?

*Ans.* 1 in work and −1 in force.

22. If the unit of force be the weight of one pound, what must be the unit of mass so that the equation $P=mf$ may still be true?

*Ans.* $g$ pounds.

## Verification of formulae by means of counting the dimensions.

**112.** Many formulae and results may be tested by means of the dimensions of the quantities involved. Suppose we have an equation between any number of physical quantities. Then the sum of the dimensions in each term of one side of the equation in mass, length, and time respectively must be equal to the corresponding sums on the other side of the equation. For suppose that the dimensions in length of one side of the equation differed from the corresponding dimensions on the other side of the equation; then on altering the unit of length the two members of the equation would be altered in different ratios and would be no longer equal; this however would be clearly absurd; for two quantities which are equal must have the same measures whatever (the same) unit is used. For example, if two sums of money are the same, their measures must be the same whether we express the amounts in pounds, shillings, or pence.

**113.** Much information may be often easily obtained by considering the dimensions of the quantities involved. Thus the time of oscillation of a simple pendulum (which consists of a mass $m$ tied by means of a light string of length $l$ to a fixed point) may be easily shewn to vary as $\sqrt{\dfrac{l}{g}}$. For, assuming the time of oscillation to be independent of the arc of oscillation, the only quantities that can appear in the answer are $m$, $l$, and $g$. Let us assume the time of oscillation to vary as $m^\alpha l^\beta g^\gamma$.

The dimensions of this quantity expressed in the usual way are
$$[M]^\alpha [L]^\beta \left\{\frac{[L]}{[T]^2}\right\}^\gamma,$$

or
$$[M]^\alpha [L]^{\beta+\gamma} [T]^{-2\gamma}.$$

Now the answer is necessarily of one dimension in time, and none in mass or length. Hence we have
$$\alpha = 0;\ \beta + \gamma = 0;\ -2\gamma = 1.$$
$$\therefore \gamma = -\tfrac{1}{2},\ \beta = \tfrac{1}{2},$$

and the time of oscillation $\propto \sqrt{\dfrac{l}{g}}$.

**114.** As another example, consider the case of a body which moves in a straight line with an acceleration $\dfrac{\mu}{d^2}$ toward a fixed point in the straight line, where $d$ is the distance of the moving body from that fixed point.

Now $\dfrac{\mu}{d^2}$ is an acceleration and must therefore be of one dimension in length and minus two in time.

$\therefore \mu$ must be of 3 dimensions in length and $-2$ in time.

We can easily shew that the time the body takes to move to the centre of force, starting from rest at a distance $a$ from it, must vary as $\sqrt{\dfrac{a^3}{\mu}}$.

For the only quantities that can appear in the answer are $\mu$, and $a$. Let us assume then that this time

$$\propto \mu^x \cdot a^y.$$

The dimensions of this latter quantity are then

$$\{[L]^3 [T]^{-2}\}^x \cdot [L]^y \text{ or } [L]^{3x+y} [T]^{-2x}.$$

Hence, as before, $-2x = 1$, and $3x + y = 0$;

$$\therefore x = -\tfrac{1}{2} \text{ and } y = \tfrac{3}{2},$$

$\therefore$ the required time $\propto \mu^{-\frac{1}{2}} a^{\frac{3}{2}}$

or as

$$\sqrt{\frac{a^3}{\mu}}.$$

**Ex. 1.** If a body be projected with velocity $V$ from a point in a horizontal plane, shew that, neglecting the resistance of the air, the range on the horizontal plane must

$$\propto \frac{V^2}{g}.$$

**Ex. 2.** From a consideration of dimensions only shew that if a body be acted on by a constant force in the direction of its motion, the space passed over in time $t$ from rest must vary as the square of the time. Would the same considerations enable us to find the space when there is an initial velocity?

**Ex. 3.** A particle moves from rest under a central force equal to $\frac{\mu}{\text{distance}}$; if $T$ be the time from rest at distance $a$ to the centre of force, shew, by the consideration of dimensions only, that

$$T \propto \frac{a}{\sqrt{\mu}}.$$

**115. Attraction units.** The law of gravitation states that every particle in nature attracts every other particle with a force which varies directly as the product of the masses of the particles, and inversely as the square of the distance between them. Hence the force of attrac-

ATTRACTION UNITS. 159

tion between two particles of masses $m_1$, $m_2$, placed at a distance $r$ apart, is $\lambda \frac{m_1 m_2}{r^2}$ where $\lambda$ is some constant quantity.

This expression for the mutual attraction of two particles can be simplified by properly choosing either the unit of mass or the unit of force.

The unit of mass can be so chosen that the constant $\lambda$ may be unity and the attraction between the two masses expressed in dynes. This may be done by considering the attraction of the earth on a particle at its surface.

Let the mass of the earth be $E$ grammes, and let the new unit of mass be $x$ grammes, so that the mass of the earth is $\frac{E}{x}$ new units of mass, and one gramme is $\frac{1}{x}$ new units of mass.

Then the attraction between the earth and one gramme is $\dfrac{\frac{E}{x} \cdot \frac{1}{x}}{R^2}$ dynes where $R$ is the radius of the earth in centimetres.

But the attraction of the earth on one gramme

$$= \text{weight of one gramme} = 981 \text{ dynes.}$$

Hence $\dfrac{\frac{E}{x} \cdot \frac{1}{x}}{R^2} = 981.$

Now $E = 6\cdot 14 \times 10^{27}$; $R = 6\cdot 37 \times 10^8$.

$$\therefore x = \sqrt{\frac{E}{981 R^2}} = 3928 \text{ approximately.}$$

Hence the unit of mass, when the attraction between two masses is $\frac{m_1 m_2}{r^2}$ dynes, is 3928 grammes approximately.

The corresponding unit of mass belonging to the Foot-Pound-Second system is $\sqrt{\frac{E}{g \cdot R^2}}$ pounds, where $E$ is the number of lbs. in the earth's mass, $R$ is the number of feet in the radius of the earth, and $g$ is equal to 32 approximately.

**116.** We can, if we wish, simplify the expression for the attraction by keeping a gramme as the unit of mass and choosing a new unit of force.

In this case the new unit of force will be the attraction between two masses, each one gramme, placed at a distance of one centimetre from one another. Hence the attraction between the earth and a gramme on its surface will be $\frac{E \cdot 1}{R^2}$ units of force. But this attraction is 981 dynes.

Hence the unit of force $= \frac{981 \cdot R^2}{E}$ dynes.

But, as in the previous article, $E = 6\cdot 14 \times 10^{27}$; $R = 6\cdot 37 \times 10^8$.
Hence we easily see that this unit of force would be $6\cdot 48 \times 10^{-8}$ dynes.

**117.** The constant $\lambda$ can be so determined that the masses are expressed in grammes and the attraction between them in dynes. In this case the attraction between the earth and a mass of one gramme on its surface is

$$\lambda \cdot \frac{E \cdot 1}{R^2} \text{ dynes.}$$

Hence
$$\lambda \cdot \frac{E}{R^2} = 981.$$

$$\therefore \lambda = \frac{981\, R^2}{E} = 6\cdot 48 \times 10^{-8}.$$

The attraction between two masses, of $m_1$ and $m_2$ grammes respectively, placed at a distance of $r$ centimetres is therefore $6\cdot 48 \times 10^{-8} \times \frac{m_1 m_2}{r^2}$ dynes.

## 118. Astronomical unit of mass.

Instead of using a pound or a gramme as the unit of mass we can simplify the expression for the attraction between two bodies by choosing as our unit mass that mass which would attract an equal mass with the corresponding unit of force; this unit of force is the force which in the new unit of mass produces unit acceleration.

Thus, using centimetre-second units, let $x$ grammes be the new unit of mass. Then, as in Art. 115,

$$\frac{x \cdot \frac{1}{x}}{R^2} = \text{weight of one gramme}$$

$$= \text{weight of } \frac{1}{x} \text{ units of mass} = \frac{1}{x} \times 981,$$

so that $x = \dfrac{E}{981 R^2} = 1{\cdot}543 \times 10^7$ approximately.

This unit is called the Astronomical Unit of Mass of the centimetre-second system, and is equal to $1{\cdot}543 \times 10^7$ grammes.

The attraction between two masses $m_1$, $m_2$, at a distance $r$ apart, is $\dfrac{m_1 m_2}{r^2}$. This attraction is, therefore, of dimensions $[M]^2 [L]^{-2}$. But a force is of dimensions $[M][L][T]^{-2}$.

Hence $\quad [M]^2 [L]^{-2} = [M] [L] [T]^{-2}$.

$$\therefore [M] = [L]^3 [T]^{-2}.$$

Hence the astronomical unit of mass is of 3 dimensions in length and $-2$ in time.

The astronomical unit of density is of dimensions $[M][L]^{-3}$ or of dimensions $[T]^{-2}$. It is therefore independent of the unit of length.

## Table of Dimensions and Values of Fundamental Quantities.

[Many of the following values are taken from Prof. Everett's *Units and Physical Constants*.]

| Physical Quantity | Dimensions in Mass | Dimensions in Space | Dimensions in Time |
|---|---|---|---|
| Volume density | 1 | −3 | |
| Surface density | 1 | −2 | |
| Line density | 1 | −1 | |
| Velocity | | 1 | −1 |
| Acceleration | | 1 | −2 |
| Force | 1 | 1 | −2 |
| Momentum | 1 | 1 | −1 |
| Impulse | 1 | 1 | −1 |
| Kinetic energy | 1 | 2 | −2 |
| Power or Rate of work | 1 | 2 | −3 |

### Values of "$g$."

| Place | Ft.-sec. units | Cm.-sec. units |
|---|---|---|
| The equator | 32·091 | 978·10 |
| Cape of Good Hope | 32·140 | 979·62 |
| Latitude 45° | 32·17 | 980·61 |
| Paris | 32·180 | 980·94 |
| London | 32·19 | 981·17 |
| North Pole | 32·252 | 983·11 |

The value of $g$ at any place is approximately given by

$$g = g_0 [1 - ·00257 \cos 2\lambda - 1·96 \times h \times 10^{-9}],$$

where $g_0$ is the value of gravity in latitude 45° at the sea level, $\lambda$ the latitude of the place and $h$ its height above the sea level in cms.

# UNITS AND PHYSICAL CONSTANTS.

| | |
|---|---|
| 1 cm. | = ·39370 inches = ·032809 feet. |
| 1 foot | = 30·4797 cms. |
| 1 gramme | = 15·432 grains = ·0022046 lb. |
| 1 lb. | = 453·59 grammes. |
| 1 dyne | = weight of $\frac{1}{981}$ gramme approx. |
| 1 poundal | = 13825 dynes. |
| 1 foot-poundal | = 421390 ergs. |
| 1 foot-pound | = 1·356 × $10^7$ ergs. |
| 1 foot-pound | = 13825 gramme-centimetres. |
| 1 erg | = 7·37 × $10^{-8}$ foot-pounds. |
| 1 Joule | = $10^7$ ergs = about $\frac{3}{4}$ foot-pound. |
| 1 Horse-power | = 7·604 × $10^6$ grms-cms. per sec. |
| | = 7·46 × $10^9$ ergs per sec. |
| | = 746 Watts. |
| 1 Force de Cheval | = 7·5 × $10^6$ grms-cms. per sec. |
| 1 Watt | = 1 Joule per sec. |
| | = $10^7$ ergs per sec. |

| | |
|---|---|
| Radius of the earth | = 6·37 × $10^8$ cms. |
| Mean density of earth | = 5·67 grammes per cm. |
| Mass of earth | = 6·14 × $10^{27}$ grammes. |
| Radius of the moon | = 1·74 × $10^8$ cms. |
| Mean density of moon | = 3·6 grammes per cm. |
| Mass of the moon | = 6·98 × $10^{25}$ grammes. |

Mean distance between centres of moon and earth
$$= 3·84 \times 10^{10} \text{ cms.}$$

Mass of the sun = 324000 × that of earth.

Mean distance of sun and earth = 1·487 × $10^{13}$ cms.
$$= 9·239 \times 10^7 \text{ miles.}$$

Velocity of a point on the equator of the earth about its axis
$$= 46510 \text{ cms. per sec.}$$

Mean velocity of the earth in its orbit
$$= 2960600 \text{ cms. per sec.}$$

## EXAMPLES. CHAPTER IV.

1. If the unit of acceleration be that of a body falling freely, the unit of velocity the velocity acquired by the body in $a$ seconds from rest, and the unit of momentum that of one pound after falling for $b$ seconds; find the units of length, time, and mass.

2. Find the units of mass, length, and time supposing that when a force equal to the weight of a gramme acts on the mass of 16 grammes the acceleration produced is the unit of acceleration, that the work done in the first four seconds is the unit of work, and that the force is doing work at unit rate when the body is moving at the rate of 90 cms. per second.

3. In a certain system of absolute units the acceleration produced by gravity in a body falling freely is denoted by 3, the kinetic energy of a 600 pound shot moving with velocity 1600 feet per second is denoted by 100, and its momentum by 10; find the units of length, mass, and time.

4. If the unit of velocity be a velocity of one mile per minute, the unit of acceleration the acceleration with which this velocity would be acquired in 5 minutes, and the unit of force equal to the weight of half a ton, find the units of length, time, and mass.

5. The velocity of a train running 60 miles per hour is denoted by 8, the resistance the train experiences and which is equal to the weight of 1600 lbs. is denoted by 10, and the number of units of work done by the engine per mile by 10. Find the units of mass, length, and time.

EXAMPLES ON CHAPTER IV. 165

6. In two different systems of units an acceleration is represented by the same number whilst a velocity is represented by numbers in the ratio 1 : 3; compare the units of time and space.

If further the momentum of a body be represented by numbers in the ratio 5 : 2 compare the units of mass.

7. The units of mass, energy, and distance being taken as fundamental, find the dimensions of the unit of time in terms of these units.

8. What is the number of lbs. in the unit of mass, if the attraction between two particles of masses $m_1$, $m_2$ respectively placed at a distance of $r$ feet be $\frac{m_1 m_2}{r^2}$ poundals, assuming the earth to be a sphere of 4000 miles radius and of mean density 5·67 times that of water?

9. Assuming the attraction between two masses $m_1$, $m_2$ to be $\lambda \frac{m_1 m_2}{r^2}$, shew that, when ft.-lb.-sec. units are used, the value of $\lambda$ is approximately $1\cdot 02 \times 10^{-9}$.

[Use the values given in the preceding question.]

10. If the unit of force be defined as the attraction between two unit masses at unit distance, find the unit of mass in pounds. [The units of space and time are taken to be one foot and one second respectively, and gravity to be due to the attraction of a sphere of 4000 miles radius and having a specific gravity of 6, that of water being 1.]

Find also the attraction of two lbs. mass which are a yard apart in terms of the weight of one pound.

11. Assuming the attraction between two masses $m_1$, $m_2$ to be $\lambda \frac{m_1 m_2}{r^2}$, when absolute units are used, find the dimensions of $\lambda$.

12. Find in dynes the attraction of gravitation of two homogeneous spheres, each of mass 10 kilogrammes, the distance between their centres being a metre; given that a quadrant of the earth supposed spherical is $10^9$ centimetres the mean density of the earth 5·67 grammes per centimetre, and the acceleration due to gravity 981 c.g.s. units.

13. Taking the value of gravity as 981 in the centimetre-sec. system and the earth's radius as $6·37 \times 10^8$ centimetres, find the earth's mass in astronomical units.

14. Shew that the attraction between the moon and the earth is about $2·169 \times 10^{25}$ dynes having given the following data; moon's distance $= 60 \times$ earth's distance; moon's radius $= 1740$ kilometres; moon's density $= 3·6$ grammes per centimetre, and acceleration due to gravity on the earth $= 982·8$ c.g.s. units.

# CHAPTER V.

### PROJECTILES.

119. In the previous chapters we have considered only motion in straight lines. In the present chapter we shall consider the motion of a particle projected into the air with any direction and velocity. We shall suppose the motion to be within such a moderate distance of the earth's surface that the acceleration due to gravity may be considered to remain sensibly constant. We shall also neglect the resistance of the air and consider the motion to be *in vacuo*; for, firstly, the law of resistance of the air to the motion of a particle is not accurately known, and, secondly, even if this law were known, the discussion would require a much larger range of knowledge of pure mathematics than the reader of the present book is supposed to possess.

120. Before proceeding further it may be advisable to state some of the properties of the curve called the **parabola** which are most useful for our purpose. We shall give no proofs. They may be found in any book on Geometrical Conics such as those by Dr Taylor, Dr Besant, or the Rev. W. H. Drew.

# PROPERTIES OF THE PARABOLA.

(1) A parabola is the path traced out by a point which moves so that its distance from a fixed point $S$ is the same as its perpendicular

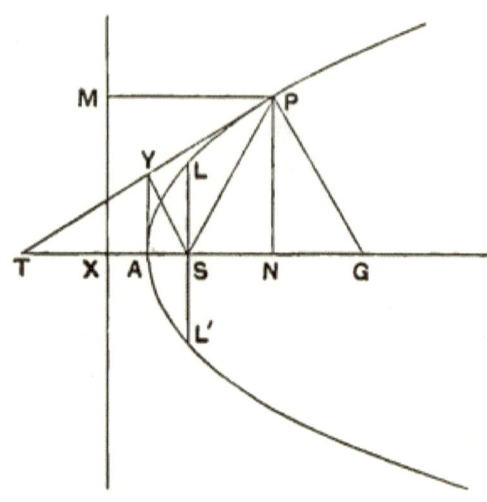

distance from a fixed straight line $XM$, so that if $P$ be any point on the parabola then $SP = PM$.

(2) The point $S$ is called the focus, $XM$ the directrix, and the line $SX$, perpendicular to $XM$, the axis of the curve. The point $A$ where the axis meets the curve is called the vertex.

(3) The double ordinate $LSL'$ through the focus perpendicular to the axis is called the latus rectum and is equal to $2SX$ or $4AS$.

(4) If $PN$ be the perpendicular from any point on the axis, then $PN^2 = 4AS \cdot AN$.

(5) The tangent $PT$ at $P$ makes equal angles with the axis and the focal distance $SP$, and therefore $SP = ST$; also $AT = AN$.

(6) If $PG$ be the normal, then $NG$ is equal to the semi-latus rectum.

(7) If $SY$ be drawn perpendicular to the tangent at any point $P$, then $Y$ lies on the tangent at the vertex $A$ and $SY^2 = AS \cdot SP$.

## PROPERTIES OF THE PARABOLA. 169

(8) A line through any point $P$ of the curve parallel to the axis is called the diameter through $P$ and any chord $QVQ'$ drawn parallel to the tangent at $P$ is an ordinate to the diameter $PV$.

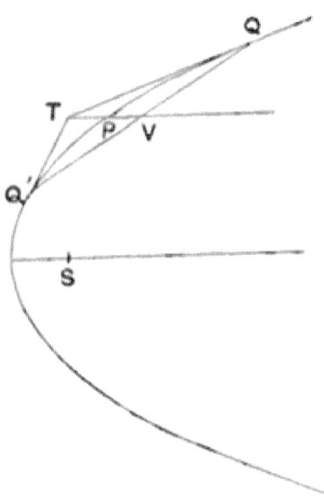

(9) If $QQ'$ be any ordinate to the diameter $PV$, then the tangents at $Q$, $Q'$ meet $PV$ in a point $T$ such that $TP = PV$, the line $QQ'$ is bisected in $V$, and $QV^2 = 4SP \cdot PV$.

Conversely if we can shew that the square of the distance $QV$, drawn from a variable point $Q$ in a fixed direction to meet a fixed line $PV$ in $V$, is proportional to the distance intercepted between $V$ and a fixed point $P$, it follows by a *reductio ad absurdum* proof that $Q$ lies on a parabola whose axis is parallel to $PV$ and whose tangent at $P$ is parallel to $QV$.

**121. *Def.*** When a particle is projected into the air the angle that the direction in which it is projected makes with the horizontal plane through the point of projection is called the **angle of projection**; the path which the particle describes is called its **trajectory**; and the distance between the point of projection and the point where the path meets any plane drawn through the point of projection is its **range** on the plane.

# 170   THE PATH OF A PROJECTILE

**122. Theorem.** *A particle is projected into the air with a given velocity and angle of projection; to shew that its path is a parabola.*

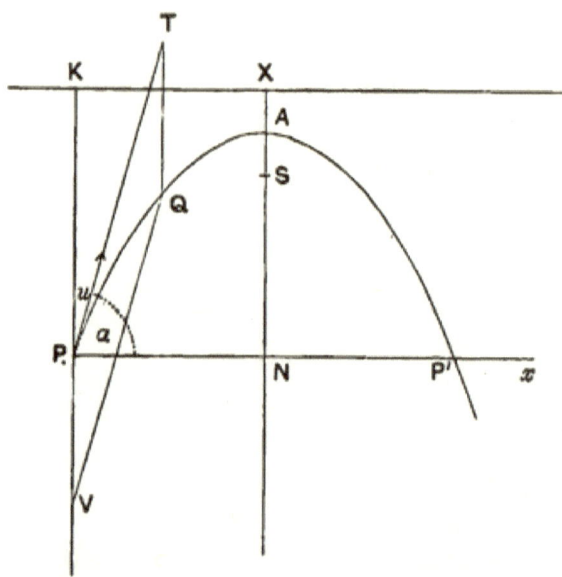

Let $P$ be the point of projection, $u$ the velocity and $a$ the angle of projection $TPx$.

The motion will take place in the vertical plane passing through $PT$; for there is nothing to take the particle out of this plane, the only force which acts on the particle, viz. its weight, being in this plane.

Consider the position of the particle at the end of any time $t$; measure off $PT$ along the direction of projection equal to $ut$. Draw $TQ$ vertically downwards and equal to $\frac{1}{2}gt^2$.

Then at the end of time $t$ the particle would, if no force acted on it, be at $T$; but its weight gives it an acceleration $g$ in a direction vertically downwards and therefore would, in time $t$, draw it through a distance $\frac{1}{2}gt^2$.

Hence, by the principle of the Physical Independence of Forces, (Art. 79), the particle is at $Q$ at the end of time $t$.

Draw $QV$ parallel to $PT$ to meet the vertical line through $P$ in $V$.

Then $QV = PT = ut$, and $PV = TQ = \frac{1}{2}gt^2$.

$\therefore$ for all values of $t$, $QV^2 = u^2t^2 = \dfrac{2u^2}{g} \cdot PV$.

Hence if a parabola be drawn passing through $P$, having $PV$ as a diameter, touching $PT$ at $P$, and such that, if $S$ be its focus, then $4SP = \dfrac{2u^2}{g}$, it follows by Art. 120 (9) that $Q$ is always on this parabola.

123. *The velocity of a projectile at any point of its path is equal, in magnitude, to the velocity that the projectile would acquire in falling freely from the directrix to that point.*

For let $KX$ be the directrix of the parabola and draw $PK$ perpendicular to it.

Then, as in the last article,
$$u^2 = 2g \cdot SP = 2g \cdot PK.$$

Hence $u$ is the velocity that would be acquired by a particle in falling through a vertical distance $PK$. (Art. 51.)

But any point on the curve may be considered as a starting point for the subsequent portion of the path; for the particle after passing through any point of its path moves in the same way as it would if projected from that point with the velocity which it then has in the direction in which it is then moving. Hence the proposition is true for every point on the path of the projectile.

*Cor.* From this proposition it follows that the height of the directrix above any point of the path depends *only* on the magnitude of the velocity at that point and not on the direction. Hence if a number of particles be projected with the same velocity from the same point in the same vertical plane but at different inclinations to the horizon, their paths have a common directrix, and the foci of their paths lie on a circle whose centre is the point from which the particles are projected.

124. Instead of considering the motion in two directions which are respectively parallel to the **direction** of motion and vertical, we shall generally find it more convenient to separately consider the motion in the horizontal and vertical directions.

The initial horizontal velocity is $u \cos \alpha$ and, since there is no force acting on the particle in the horizontal direction, there is no change in the horizontal velocity.

The initial vertical velocity is $u \sin \alpha$ and the vertical acceleration is $-g$. Hence the *vertical* motion of the particle is the same as that of a particle starting with initial velocity $u \sin \alpha$ and moving with acceleration $-g$, and the *horizontal* motion of the particle is the same as that of a particle moving uniformly with constant velocity $u \cos \alpha$.

The motion is the same as that of a particle projected with velocity $u \sin \alpha$ inside a smooth vertical tube of small bore, whilst the tube at the same time moves parallel to itself with constant velocity $u \cos \alpha$.

**125. *To find the greatest height attained by a projectile.***

Let $A$ be the highest point of the path, and let the time of describing the arc $PA$ be $T$.

Then $T$ is the time in which the initial vertical velocity is destroyed by gravity.

Hence, by Art. 41, $0 = u \sin \alpha - gT$.

$$\therefore T = \frac{u \sin \alpha}{g},$$

giving the time to the highest point of the path. Also, by the same article, $0 = u^2 \sin^2 \alpha - 2g \cdot AN$.

$$\therefore AN = \frac{u^2 \sin^2 \alpha}{2g},$$ giving the *greatest height attained*.

**126. *To find the range on the horizontal plane through the point of projection and to determine when it is greatest.***

When the particle arrives at $P'$ the distance it has described in a vertical direction is zero. Hence, if $t$ be the time of flight, we have

$$0 = u \sin \alpha t - \tfrac{1}{2} g t^2.$$

$$\therefore t = \frac{2u \sin \alpha}{g}.$$

But during this time $t$ the horizontal velocity has remained constant and equal to $u \cos \alpha$.

$\therefore PP' =$ horizontal distance described in time $t$

$$= u \cos \alpha t = \frac{2u^2 \sin \alpha \cos \alpha}{g}.$$

Hence the range is equal to twice the product of the initial vertical and horizontal velocities divided by $g$.

Again with a *given* velocity of projection the range is a maximum when $\sin 2\alpha$ is greatest and then $\alpha = 45°$.

# TWO DIRECTIONS OF PROJECTION.

**127.** From the preceding article it may be shewn that, in general, there are for a given range two directions of projection which are equally inclined to the direction of greatest range.

For let $\theta$ be the required angle of projection when the range and velocity of projection are $K$ and $u$ respectively.

Then
$$\frac{u^2 \sin 2\theta}{g} = K.$$

$$\therefore \sin 2\theta = \frac{gK}{u^2}.$$

Now, if $gK$ be not greater than $u^2$, there are two values of $2\theta$, both less than $180°$, for each of which the sine has the same magnitude, and these two values are supplementary. Hence, if $\theta_1$, $\theta_2$ be the corresponding values of $\theta$, we have

$$\sin 2\theta_1 = \frac{gK}{u^2} = \sin 2\theta_2.$$

$$\therefore 2\theta_1 = \pi - 2\theta_2.$$

$$\therefore \frac{\pi}{4} - \theta_1 = \theta_2 - \frac{\pi}{4}.$$

Hence the two directions of projection $PT_1$, $PT_2$ are

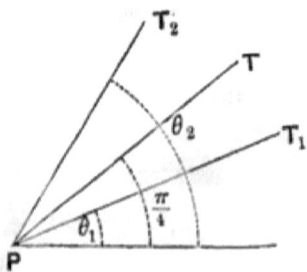

equally inclined to the direction $PT$ which gives the maximum range.

**128.** *To find the latus-rectum of the path described.*

Consider the motion at the highest point $A$ of the path. The vertical velocity here is just zero and the particle is moving horizontally with the constant horizontal velocity $u \cos a$.

But the velocity at $A$ is that due to a fall through the distance $AX$.

$$\therefore u^2 \cos^2 a = 2g \cdot AX.$$

$$\therefore \text{latus-rectum} = 2SX = 4AX = \frac{2u^2 \cos^2 a}{g}$$

= twice the square of the horizontal velocity divided by $g$.

It will be noted that the latus-rectum, and therefore the *size* of the parabola, is independent of the initial vertical velocity and depends only on the horizontal velocity.

*Cor.* The height of the focus above the horizontal line through $P$

$$= AN - AS = \frac{u^2 \sin^2 a}{2g} - \frac{u^2 \cos^2 a}{2g} = -\frac{u^2}{2g} \cos 2a.$$

Hence, if $a$ be less than $45°$, the focus of the path is situated *below* the horizontal line drawn through the point of projection.

**129.** *To find the velocity and direction of motion after a given time.*

Let $U$ be the velocity and $\psi$ the angle which the direction of motion at the end of time $t$ makes with the horizontal.

Then $U\cos\psi$ = horizontal velocity at this time
$$= u\cos\alpha, \text{ the constant horizontal velocity.}$$
And $U\sin\psi$ = the vertical velocity at this time
$$= u\sin\alpha - gt.$$

Hence, by squaring and adding,
$$U^2 = u^2 - 2ugt\sin\alpha + g^2t^2,$$
and, by division, $\tan\psi = \dfrac{u\sin\alpha - gt}{u\cos\alpha}.$

**Cor.** From this article we can at once deduce the important proposition in Art. 123. For the height of the particle at time $t$ above its original starting point $= u\sin\alpha \cdot t - \frac{1}{2}gt^2$.

$\therefore$ its depth below the directrix at time $t = \dfrac{u^2}{2g} - (u\sin\alpha \cdot t - \frac{1}{2}gt^2)$,

and the square of the velocity acquired in falling through this distance
$$= 2g \times \left[\frac{u^2}{2g} - u\sin\alpha \cdot t + \frac{1}{2}gt^2\right] = u^2 - 2ugt\sin\alpha + g^2t^2 = U^2.$$

**130.** *Given the velocity and direction of projection, to construct the focus and directrix of the path.*

Let $P$ be the point of projection and $PT$ the direction of projection, so that $PT$ is the tangent to the path at $P$.

Draw $PK$ vertical and equal to $\dfrac{u^2}{2g}$, so that $K$ is a point on the directrix and a horizontal line through $K$ is the directrix.

Draw $PS$ on the other side of $PT$ from $PK$ so that the angle $TPS$ is equal to the angle $TPK$. Then, by Art. 120 (5), it follows that the focus must lie on the line $PS$.

Make $PS$ equal to $PK$; then $S$ is clearly the focus.

## EQUATION TO PATH.

**131.** *To find the equation to the path of* **the projectile.**

Take as axes of coordinates the horizontal and vertical lines through **the** point of **projection,** $P$. Let $Q$ be the position of the particle at the end **of** time $t$, and let $(x, y)$ be its coordinates.

Then $y$ **is the** vertical distance described in time $t$ by a particle starting with velocity $u \sin \alpha$ **and** moving with acceleration $-g$.

$$\therefore y = u \sin \alpha \cdot t - \tfrac{1}{2} g t^2.$$

So $\qquad x = u \cos \alpha \cdot t.$

Hence, eliminating $t$, we have

$$y = x \tan \alpha - \frac{g}{2} \frac{x^2}{u^2 \cos^2 \alpha}.$$

This equation can be expressed in the form

$$\left(x - \frac{u^2}{g} \sin \alpha \cos \alpha\right)^2 = -\frac{2 u^2 \cos^2 \alpha}{g} \left(y - \frac{u^2 \sin^2 \alpha}{2g}\right).$$

Hence **the** path is a **parabola whose vertex is the point** $\left(\dfrac{u^2}{g} \sin \alpha \cos \alpha, \dfrac{u^2 \sin^2 \alpha}{2g}\right)$, whose latus-rectum is $\dfrac{2 u^2 \cos^2 \alpha}{g}$, and whose **axis is** vertical and concavity turned downwards.

### EXAMPLES.

1. *A bullet is projected with an initial velocity* **of 1280** *feet per second at an angle* $\sin^{-1} \tfrac{1}{8}$ *with the horizon; find the range on the horizontal plane, the time of flight, and the greatest height attained.*

The initial horizontal velocity

$$= 1280 \cos (\sin^{-1} \tfrac{1}{8}) = 1280 \cdot \frac{\sqrt{63}}{8} = 160 \sqrt{63} \text{ feet per second,}$$

and the initial vertical velocity $= 1280 \times \tfrac{1}{8} = 160$ feet per second.

L. D.

Let $T=$ time to the highest point of the path. Then $T$ is the time in which a particle starting with a velocity of 160 feet per second, and moving with acceleration $-g$, comes to rest.

$$\therefore T = \frac{160}{g} = 5 \text{ seconds.}$$

$\therefore$ greatest height attained

= space described in 5 seconds by a particle starting with velocity 160 feet per second and moving with acceleration $-g$

$$= 160\,T - \tfrac{1}{2}gT^2 = 160 \cdot 5 - \tfrac{1}{2} \cdot 32 \cdot 5^2 = 400 \text{ feet.}$$

Also the range = distance described in 10 seconds by a particle moving uniformly with a velocity of $160\sqrt{63}$ feet per second

$$= 1600\sqrt{63} = 12700 \text{ feet approximately.}$$

2. *A man throws a stone with a velocity of 64 feet per second to hit a small object on the top of a wall whose height is 19 feet and whose horizontal distance from the man is 48 feet; find the direction of projection and the velocity and direction of the stone when it hits the object.*

Let $\theta$ be the angle of projection and $T$ the time that elapses before the object is struck.

By considering separately the horizontal and vertical motion we have

$$48 = 64 \cos\theta \cdot T, \text{ and } 19 = 64 \sin\theta \cdot T - \tfrac{1}{2}gT^2.$$

$\therefore$ eliminating $T$,

$$19 = 48\tan\theta - \tfrac{32}{2}\left(\frac{3}{4\cos\theta}\right)^2 = 48\tan\theta - 9(1+\tan^2\theta).$$

Hence $\tan\theta = \tfrac{2}{3}$ or $\tfrac{14}{3}$, giving the two possible directions of projection.

Also the vertical and horizontal components of the velocity of the stone when it hits the object are $64\sin\theta - gT$, i.e. $64\sin\theta - 24\sec\theta$, and $64\cos\theta$ respectively.

Taking the first value of $\theta$ these are respectively $\tfrac{24}{13}\sqrt{13}$ and $\tfrac{192}{13}\sqrt{13}$ feet per second, whence may be obtained the required velocity and direction of the stone. Similarly taking the second value of $\theta$ the velocity corresponding to the alternative path may be found.

3. *From the top of a cliff, 80 feet high, a stone is thrown so that it starts with a velocity of 128 feet per second at an angle of $30°$ with the horizontal; find where it hits the ground at the bottom of the cliff.*

The initial vertical velocity is $128\sin 30°$ or 64 feet per second, and the initial horizontal velocity is $64\sqrt{3}$ feet per second.

EXAMPLES. 179

Let $T$ be the time that elapses before the stone hits the ground. Then $T$ is the time in which a **stone** projected with velocity 64 and moving with **acceleration** $-g$ describes a distance $-80$ feet.

$$\therefore -80 = 64T - \tfrac{1}{2}gT^2, \text{ and therefore } T = 5.$$

During this time the horizontal velocity remains constant, and hence the distance of the **point** where the **stone** hits the ground from the foot of the cliff is $320\sqrt{3}$ feet.

4. A shot leaves a gun at the rate of 500 feet per second; calculate the greatest distance to which it could be projected and also the height it would rise.

*Ans.* $7812\tfrac{1}{2}$ feet; $1953\tfrac{1}{8}$ feet.

5. Find the velocity **and direction of** projection **of a shot which** passes in a horizontal direction just over the **top of a wall which is** 50 yards off and 75 feet high.

*Ans.* $40\sqrt{6}$ feet per second at an inclination of 45° to the horizon.

6. A projectile **fired** horizontally from a height of 9 feet from the ground reaches the ground at a horizontal distance of 1000 feet. What is its initial velocity?

*Ans.* $1333\tfrac{1}{3}$ feet per second.

7. A particle is projected horizontally from the top **of a tower, 100** feet high, and **the focus of the parabola** which it describes is in **the** horizontal plane through **the foot of** the tower; find the **velocity of** projection.

*Ans.* 80 feet per second.

8. A particle is projected with velocity $2\sqrt{ag}$ so that it just clears two walls, of equal height $a$, which are at a distance $2a$ from each other. Shew that the latus-rectum of the path is $2a$ and that the time of passing between the walls is $2\sqrt{\dfrac{a}{g}}$.

9. A shot is fired from a gun on **the top of a cliff, 400 feet high, with** a velocity of 768 feet per second at **an** elevation **of 30°**. Find the horizontal distance from the vertical line through the gun of the point where the shot strikes the water.

*Ans.* $3200\sqrt{3}$ yards.

180   EXAMPLES.

10. Shew that four times the square of the number of seconds in the time of flight in the range on a horizontal plane equals the height in feet of the highest point of the trajectory.

11. Find the angle of projection when the range on the horizontal plane through the point of projection is equal to the height to which the velocity of projection is due.

*Ans.* 15° or 75°.

12. If the maximum height of a projectile above a horizontal plane passing through the point of projection be $h$ and $a$ be the angle of projection, find the interval between the instants at which the height of the projectile is $h \sin^2 a$.

*Ans.* $2\sqrt{\dfrac{2h}{g}} \cos a.$

13. In a trajectory find the time that elapses before the particle is at the end of the latus-rectum.

*Ans.* $\dfrac{u}{g} (\sin a \pm \cos a).$

14. Find the direction in which a rifle must be pointed so that the bullet may strike a body let fall from a balloon at the instant of firing; find also the point where the bullet meets the body, supposing the balloon to be 220 yards high, the angle of its elevation from the position of the rifleman to be 30°, and the velocity of projection of the bullet to be two miles per second.

*Ans.* The rifle must be pointed at the balloon; the bullet will strike the body when it is 3 inches below the point where it left the balloon.

15. A stone is thrown in such a manner that it would just hit a bird at the top of a tree and afterwards reach a height double that of the tree; if, at the moment of throwing the stone, the bird fly away horizontally, shew that, notwithstanding this, the stone will hit the bird if its horizontal velocity be to that of the bird as $\sqrt{2}+1 : 2$.

16. If $t$ be the time in which a projectile reaches a point $P$ of its path and $t'$ be the time from $P$ till it strikes the horizontal plane through the point of projection, shew that the height of $P$ above the plane is $\tfrac{1}{2} g t t'$.

17. If at any point of a parabolic path the velocity be $u$ and the inclination to the horizon be $\theta$, shew that the particle is moving at right angles to this direction after a time $\dfrac{u}{g \sin \theta}$.

18. In any trajectory shew that the component of the velocity **perpendicular to the focal radius vector at any** point is **constant**.

19. A particle is projected so as **to** enter in the direction of **its** length a small straight **tube** of small bore fixed **at an** angle of 45° to **the horizon** and to pass out at the other end of **the** tube; shew that the latera recta of the paths which the particle describes before entering and leaving the **tube** differ by $\sqrt{2}$ **times** the length of the tube.

132. **Range on an inclined plane.** *From a point on a plane, which is inclined at **an angle** $\beta$ **to the** horizon, a particle is projected with a velocity* u *at an angle* $\alpha$ *with **the** horizontal in a plane passing through the normal to the inclined plane and the line of greatest slope,* **to find the** *range on the inclined plane.*

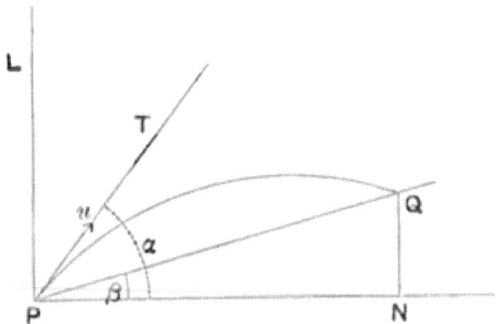

Let $PQ$ be the range on the inclined plane, $PT$ the direction of projection, and $QN$ the perpendicular on the horizontal plane **through** $P$.

**The initial component of the** velocity perpendicular **to** $PQ$ is $u \sin(\alpha - \beta)$, and the acceleration in this **direction** is $-g \cos \beta$.

Let $T$ be the time which the particle takes to go from $P$ to $Q$. Then in time $T$ the space described **in a** direction perpendicular to $PQ$ is zero.

Hence $0 = u\sin(\alpha - \beta).T - \tfrac{1}{2}g\cos\beta.T^2$, and therefore

$$T = \frac{2u}{g}\frac{\sin(\alpha - \beta)}{\cos\beta}.$$

During this time the horizontal velocity $u\cos\alpha$ remains unaltered; hence $PN = u\cos\alpha.T$.

$\therefore$ the range $PQ = \dfrac{PN}{\cos\beta} = \dfrac{u\cos\alpha}{\cos\beta}.T = \dfrac{2u^2\cos\alpha\sin(\alpha-\beta)}{g\cos^2\beta}$.

**133. Maximum range.** *To find the direction of projection which gives the maximum range on the inclined plane, and to shew that for any given range there are two directions of projection which are equally inclined to the direction for maximum range.*

From the preceding article the range

$$= \frac{2u^2}{g}\frac{\cos\alpha\sin(\alpha-\beta)}{\cos^2\beta} = \frac{u^2}{g\cos^2\beta}\{\sin(2\alpha-\beta) - \sin\beta\}\ldots\text{(i)}.$$

Now $u$ and $\beta$ are given; hence the range is a maximum when $\sin(2\alpha - \beta)$ is greatest or when $2\alpha - \beta = \dfrac{\pi}{2}$.

In this case $\alpha - \beta = \dfrac{\pi}{2} - \alpha$, and therefore the required direction of projection, $PT$, bisects the angle between the vertical, $PL$, and the inclined plane.

Hence the maximum range

$$= \frac{u^2}{g\cos^2\beta}(1 - \sin\beta) = \frac{u^2}{g(1+\sin\beta)}.$$

When the range is given let $\theta$ be the required angle of projection. Then, by (i), $\sin(2\theta - \beta)$ is given. Now there are two values of $2\theta - \beta$, each less than $180°$, for each of which the sine has the same magnitude, and these two values are supplementary. Hence if $\theta_1$, $\theta_2$, be the corresponding values of $\theta$ we have

$$\sin(2\theta_1 - \beta) = \sin(2\theta_2 - \beta).$$
$$\therefore 2\theta_1 - \beta = \pi - (2\theta_2 - \beta).$$
$$\therefore \theta_1 - \tfrac{1}{2}\left(\frac{\pi}{2} + \beta\right) = \tfrac{1}{2}\left(\frac{\pi}{2} + \beta\right) - \theta_2.$$

But $\tfrac{1}{2}\left(\dfrac{\pi}{2} + \beta\right)$ has been shewn to be the elevation for the maximum range.

Hence the two directions of projection for a given range are equally inclined to the direction of maximum range.

**134. Motion upon an inclined plane.** *A particle moves upon a smooth plane which is inclined at an angle $\beta$ to the horizon, being projected from a point in the plane with velocity* u *in a direction inclined at an angle* α *to the intersection of the inclined plane with a horizontal plane; to find the motion.*

Resolve the acceleration due to gravity into two components; one, $g \sin \beta$, in the direction of the line of greatest slope and the other, $g \cos \beta$, perpendicular to the inclined plane. The latter acceleration is destroyed by the reaction of the plane.

The particle therefore moves on the inclined plane with an acceleration $g \sin \beta$ parallel to the line of greatest slope.

Hence the investigation of the motion is the same as that in Arts. 122—131, if we substitute "$g \sin \beta$" for "$g$" and instead of "vertical distances" read "distances measured on the inclined plane parallel to the line of greatest slope".

## EXAMPLES.

1. A plane is inclined at 30° to the horizon; from its foot a particle is projected with a velocity of 600 feet per second in a direction inclined at an angle of 60° to the horizon; find the range on the inclined plane and the time of flight.

    *Ans.* 2500 yards; $25\frac{\sqrt{3}}{2}$ seconds.

2. The greatest range of a particle, projected with a certain velocity, on a horizontal plane is 5000 yards; find its greatest range on an inclined plane whose inclination is 45°.

    *Ans.* 2927 yards approximately.

3. A hill is inclined at an angle of 30° to the horizon; from a point on the hill one projectile is projected up the hill and another down; the angle of projection in each case is 45° with the horizon; shew that the range of one projectile is nearly $3\frac{2}{3}$ that of the other.

4. A particle is projected from a point on an inclined plane in a direction making an angle of 60° with the horizon; if the range on the plane be equal to the distance through which another particle would fall from rest during the time of flight of the first particle, find the inclination of the plane to the horizon.

    *Ans.* 30°.

5. From a point in a given inclined plane two bodies are projected with the same velocity in the same vertical plane at right angles to one another; shew that the difference of the ranges is constant.

6. The angular elevation of an enemy's position on a hill $h$ feet high is $\beta$; shew that in order to shell it the initial velocity of the projectile must not be less than $\sqrt{gh(1+\operatorname{cosec}\beta)}$.

7. Shew that the greatest range on an inclined plane through the point of projection is equal to the distance through which the particle could fall freely during its time of flight.

## GEOMETRICAL CONSTRUCTION.   185

135. *Geometrical construction for* **the path of a particle projected with a given velocity so as to have a given range on an inclined plane passing through the point of** *projection.*

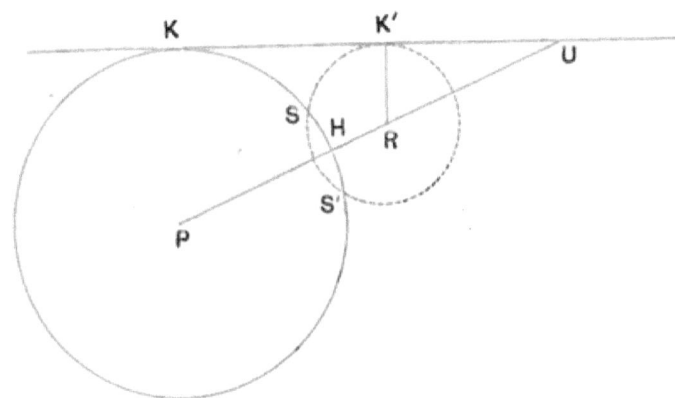

Let $P$ be the point of projection, and $u$ the velocity of projection. Draw $PK$ vertical and equal to $\dfrac{u^2}{2g}$; then $KK'$, the horizontal line through $K$, is the directrix of all the paths [Art. 123] and the foci of all the paths lie on a circle whose centre is $P$ and radius $PK$. Let this circle meet $PU$ in $H$.

To find the focus of the path corresponding to any range $PR$ we must find a point $S$ on this circle such that $RS$ may be equal to the perpendicular $RK'$ drawn to the directrix. A circle whose centre is $R$ and distance $RK'$ cuts the circle of foci in two points $S$, $S'$ on opposite sides of $PR$. Hence there are two paths and the corresponding directions of projection from $P$ bisect the angles $KPS$, $KPS'$.

It is plain that the farther $R$ is from $P$, i.e. the

greater the range, the more nearly do $S$, $S'$ approach, and the range will be a maximum when $S$ and $S'$ coincide at $H$.

Hence the direction of projection for the maximum range bisects the angle $KPU$ which is the angle between the vertical and the inclined plane; also the focus of the trajectory of maximum range on an inclined plane lies in the plane.

Again since $PS$, $PS'$ are equally inclined to $PH$ it follows that the bisectors of the angles $KPS$, $KPS'$ are equally inclined to the bisector of the angle $KPH$.

Hence the two directions of projection for any given range on the inclined plane are equally inclined to the direction for maximum range.

136. **Envelope of paths.** *To shew that the envelope of all the paths described by particles projected from a given point with the same velocity is another parabola.*

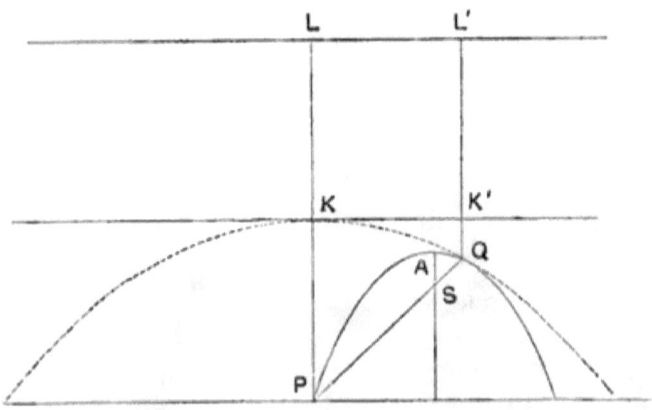

Let $P$ be the point of projection; $h$ the height to which the velocity of projection is due. Draw $PK$

vertical and equal to $h$. Then the horizontal line $KK'$ is the common directrix of all the paths.

Let $PAQ$ be any one of the paths and let $PSQ$ be its focal chord through $P$.

Produce $PK$ to $L$ making $KL$ equal to $PK$ or $h$ and draw $LL'$ horizontal.

Draw $QK'L'$ a vertical line as in the figure.

Then $PQ = PS + SQ = PK + QK' = K'L' + QK' = QL'$.

∴ $Q$ always lies on a parabola whose focus is at $P$ and of which $LL'$ is the directrix and therefore $K$ the vertex. Let this parabola be the dotted curve.

Since the line $QSP$ passes through the foci of both parabolas, the tangent to each parabola at $Q$ is the same. [Art. 120 (5).]

Hence the dotted parabola touches the original parabola at $Q$. Similarly it touches all the other paths. Hence it is their envelope.

*The envelope of all the paths is therefore a parabola whose focus is at the point of projection and whose latus-rectum is* 4h, *where* h *is the height to which the velocity of projection is due.*

137. The envelope of the paths can be easily found by analytical methods.

For the equation of the path is, by Art. 131,

$$y = x \tan a - \frac{x^2}{4h \cos^2 a}, \text{ since } u^2 = 2gh,$$

or

$$y = x \tan a - \frac{x^2}{4h}(1 + \tan^2 a),$$

or

$$\frac{x^2}{4h} \tan^2 a - x \tan a + \frac{x^2}{4h} + y = 0.$$

The equation to the envelope of this curve is (C. Smith's *Conic Sections*, Art. 237),

$$x^2 = 4 \cdot \frac{x^2}{4h}\left(\frac{x^2}{4h} + y\right),$$

or
$$x^2 = -4h(y - h).$$

This is a parabola whose focus is the point $P$, whose vertex is $K$, whose latus-rectum is $4h$, and the concavity of which is turned downwards.

**138.** Since the enveloping parabola touches externally all the paths, it follows that the **maximum range** in any given direction is obtained by finding where this direction meets the enveloping parabola; for no point *outside* the enveloping parabola can be reached by a particle starting with the given velocity of projection.

**Ex. 1.** *To find the maximum range on a plane through the point of projection inclined at an angle a to the horizon.*

Taking the figure of Art. 136, let $PQ$, the given inclined plane, meet the enveloping parabola in $Q$. Let $QM$ be the perpendicular on $PK$.

Then $2h = QL' + PM = PQ + PQ \cdot \sin a$.

$\therefore$ the greatest range $= PQ = \dfrac{2h}{1 + \sin a} = \dfrac{u^2}{g(1 + \sin a)}$.

**Ex. 2.** *To find the maximum range on a plane inclined at an angle a to the horizon and such that the perpendicular distance of the point of projection from it is d.*

Let the plane meet the enveloping parabola in $R$; draw $RN$ perpendicular to $PK$ and $PY$ perpendicular to the inclined plane. Then $RY$ is the maximum range required. Let it be $x$.

Then we have $RN^2 = 4h \cdot KN$,

$\therefore (x \cos a + d \sin a)^2 = 4h(h + d \cos a - x \sin a)$.

Solving this equation we obtain

$$x = \frac{2\sqrt{h(h + d \cos a)} - \sin a(2h + d \cos a)}{\cos^2 a},$$

which is the required maximum range.

## MISCELLANEOUS EXAMPLES.

**Ex. 1.** *A particle **is** projected at an angle $\alpha$ with the horizontal from the foot of a plane, **whose** inclination to the horizon **is** $\beta$; **shew** that it will strike the plane at right angles if* $\cot\beta = 2\tan(\alpha-\beta)$.

Let $u$ be the velocity of projection so that $u\cos(\alpha-\beta)$ and $u\sin(\alpha-\beta)$ are the initial **velocities** respectively parallel and perpendicular to the inclined plane.

The accelerations in these two directions are $-g\sin\beta$ and $-g\cos\beta$.

Then, **as** in Art. 132, the **time,** $T$, **that elapses** before the particle reaches the plane again is $\dfrac{2u\sin(\alpha-\beta)}{g\cos\beta}$.

**If the** direction of motion at the instant when the particle hits the plane **be** perpendicular to the **plane,** then the velocity **at** that instant parallel to the plane **is zero.**

Hence $\qquad u\cos(\alpha-\beta) - g\sin\beta\, T = 0.$

$$\therefore \quad \frac{u\cos(\alpha-\beta)}{g\sin\beta} = T = \frac{2u\sin(\alpha-\beta)}{g\cos\beta}.$$

$$\therefore \quad \cot\beta = 2\tan(\alpha-\beta).$$

**Ex. 2.** *Particles slide from rest down smooth **chords of a vertical** circle, starting from its highest point, and then move freely; shew **that** the locus of their foci **is** a circle of half the size of the given circle and that their paths bisect the vertical radius.*

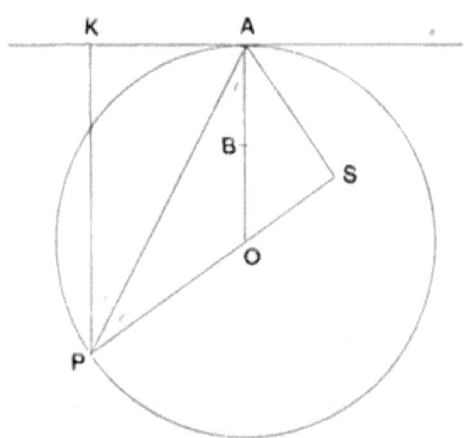

Let $A$ be the highest point of the circle, $O$ its centre, and $AP$ any chord; draw $PK$ perpendicular to $AK$. Then since (Art. 123) the velocity at $P$ is that due to the vertical depth of $P$ below $A$, it follows that $AK$ is the directrix of the path described by the particle after leaving $P$.

Also $\angle OPA = \angle OAP = \angle APK$, and $AP$ is the tangent at $P$ to the parabola described.

∴ the focus $S$ lies on $PO$.

Also $PS = PK$. Therefore the two triangles $APK$, $APS$ are equal. Hence $ASP$ is a right angle, and therefore $S$ lies on a circle on $AO$ as diameter.

Again if $B$ be the middle point of $AO$, it is the centre of this circle; therefore $SB = BA$, and $B$ is therefore a point on the parabola a portion of which is described by the particle after leaving $P$.

**Ex. 3.** *A shot of* m *pounds is fired from a gun of* M *pounds which is placed on a smooth horizontal plane and elevated at an angle* α. *Shew that, if the muzzle velocity of the shot be* V, *the range will be*

$$2\frac{V^2}{g} \frac{\left(1+\frac{m}{M}\right)\tan\alpha}{1+\left(1+\frac{m}{M}\right)^2 \tan^2\alpha}.$$

Let $u$ be the velocity communicated to the shot along the barrel of the gun by the explosion of the gunpowder, so that the impulse $I$ of the force exerted by the expanding gases is $mu$. An equal and opposite impulse is communicated to the gun; let this impulse be resolved into two, $I\cos\alpha$ horizontally, and $I\sin\alpha$ vertically.

The latter is counterbalanced by the increased pressure on the horizontal plane.

Now the shot, whilst inside the barrel, has, in addition to its velocity $u$ in the direction of the barrel, the same horizontal velocity, $U$, that the gun possesses, so that we have

$$(M+m)U = I\cos\alpha = mu\cos\alpha \quad \ldots\ldots\ldots\ldots\ldots\ldots(1).$$

If $V$ be the resultant velocity, and $\beta$ the elevation, of the shot as it leaves the barrel, we have

$$V\cos\beta = u\cos\alpha - U, \text{ and } V\sin\beta = u\sin\alpha.$$

$$\therefore \tan\beta = \frac{u\sin\alpha}{u\cos\alpha - U} = \left(1 + \frac{m}{M}\right)\tan\alpha, \text{ by substituting for } U \text{ from (1).}$$

$$\therefore \text{range} = \frac{2V^2 \sin\beta \cos\beta}{g} = \frac{2V^2}{g}\frac{\tan\beta}{1+\tan^2\beta}$$

$$= \frac{2V^2}{g} \frac{\left(1+\dfrac{m}{M}\right)\tan\alpha}{1+\left(1+\dfrac{m}{M}\right)^2 \tan^2\alpha}.$$

## EXAMPLES. CHAPTER V.

1. If $v_1$, $v_2$ be the velocities of a projectile at the ends of a focal chord of its path and $V$ the horizontal velocity, shew that

$$\frac{1}{v_1^2} + \frac{1}{v_2^2} = \frac{1}{V^2}.$$

2. A particle is projected under gravity with velocity $u$ at an angle $\alpha$ with the vertical; shew that the length of the focal chord through the point of projection is $\dfrac{u^2}{2g\cos^2\alpha}$.

3. Shew that the velocity at any point of a trajectory is equal to that acquired by a body in falling freely through a vertical distance equal to one quarter of the focal chord parallel to the tangent at the point.

4. If $\alpha$ be the angle between the tangents at the extremities of any arc of a parabolic path, $v$, $v'$ the velocities at these extremities and $u$ the velocity at the vertex of the path, shew that the time of describing the arc is $\dfrac{vv'\sin\alpha}{gu}$.

5. Shew that the velocity at any point of the path of a projectile is proportional to the length of the normal at that point; hence shew that the velocity is compounded of two equal velocities, one in a horizontal direction, and the other in a direction perpendicular to the focal distance of the point.

6. On the moon there seems to be no atmosphere and gravity there is about one sixth of that on the earth. What space of country would be commanded by the guns of a lunar fort able to project shot with a velocity of 1600 feet per second?

7. Shew that the angular velocity of a projectile about the focus of its path varies inversely as the distance of the projectile from the focus.

8. Find the charge of powder required to send a 68 lb. shot, with an elevation of $15°$, to a range of 3000 yards, given that the velocity communicated to the same shot by a charge of 10 lbs. is 1600 feet per second.

9. A coach-wheel rolling with given velocity throws off small portions of dust at a tangent to its circumference; find the greatest height from the ground to which any dust will rise.

10. The radii of the front and hind wheels of a carriage are $a$ and $b$, and $c$ is the distance between the axle-trees; a particle of dust driven from the highest point of the hind-wheel is observed to alight on the highest point of the front wheel. Shew that the velocity of the carriage is

$$\sqrt{\frac{(c+b-a)(c+a-b)}{4(b-a)}} g.$$

11. Shew that a speck of mud thrown from the top of a hansom cab-wheel of diameter $d$ feet moving with a velocity of $v$ feet per second will, when it strikes the ground, be at a distance $v\sqrt{d/g}$ in front of the position then occupied by the point of contact of the wheel and ground.

Shew also that the mud will not clear the wheel unless $v$ be

$$> \frac{\sqrt{gd}}{2}.$$

12. Three bodies are projected simultaneously from the same point in the same vertical plane, one vertically, another at an elevation of 30°, and the third horizontally; if their velocities be in the ratio $1 : 1 : \sqrt{3}$, shew that they are always in a straight line.

How does this straight line move in space?

13. Two particles are projected simultaneously, one with velocity $V$ up a smooth plane inclined at an angle of 30° to the horizon, and the other with a velocity $\dfrac{2V}{\sqrt{3}}$ at an elevation of 60°. Shew that the particles will be relatively at rest at the end of $\dfrac{2V}{3g}$ seconds from the instant of projection.

14. A particle, projected with velocity $u$, strikes at right angles a plane through the point of projection inclined at an angle $\beta$ to the horizon. Shew that the height of the point struck above the horizontal plane through the point of projection is $\dfrac{2u^2}{g} \dfrac{\sin^2 \beta}{1 + 3\sin^2 \beta}$, that the time of flight is $\dfrac{2u}{g\sqrt{1 + 3\sin^2 \beta}}$, and that the range on a horizontal plane through the point of projection would be $\dfrac{u^2 \sin 2\beta}{g} \dfrac{1 + \sin^2 \beta}{1 + 3\sin^2 \beta}$.

15. Two inclined planes intersect in a horizontal line, their inclinations to the horizon being $\alpha$ and $\beta$; if a particle be projected from a point in the former at right angles to it so that it strikes the latter plane at right angles, shew that its velocity of projection is $\sin \beta \sqrt{\dfrac{2ga}{\sin \alpha - \sin \beta \cos (\alpha + \beta)}}$, where $a$ is the distance of the point of projection from the intersection of the planes.

16. Two inclined planes of equal height $h$ and inclination $a$ are placed back to back. A ball projected along the surface of the plane at an angle $\beta$ with the horizontal flies over the top and falls at the foot of the other plane; shew that the velocity of projection is

$$\tfrac{1}{2}\sqrt{gh}\,\operatorname{cosec}\beta\,\sqrt{8+\operatorname{cosec}^2 a}.$$

17. A particle is projected from a point in the lowest line of a plane which is inclined at an angle $a$ to the horizon with velocity $V$ at an angle $\beta$ to the plane, so that the plane containing the direction of $V$ and the normal to the plane cuts the inclined plane in a line making an angle $\gamma$ with the line of greatest slope. Shew that, if the inclined plane be smooth and inelastic, the particle will describe two parabolas before it again reaches the lowest line of the plane and that the time in this second parabola is

$$\frac{4V}{g\sin 2a}\left\{\cos(a+\beta)\cos^2\frac{\gamma}{2}-\cos(a-\beta)\sin^2\frac{\gamma}{2}\right\}.$$

18. A particle is projected so as to just graze the four upper corners of a regular hexagon, whose side is $a$, placed vertically with one side on a table. Shew that the range on the table is $a\sqrt{7}$ and that the square of the time of flight is

$$\frac{28a}{g\sqrt{3}}.$$

19. If the velocity with which a bullet issues from a rifle be equal to that due to a fall from a height $h$, shew that, if the rifle be pointed directly at a point at a height $2h$ above a target, a small alteration in the elevation will not affect the position of the place where the bullet strikes the target.

20. The barrel of a rifle sighted to hit the centre of the bull's eye, which is at the same height as the muzzle and distant $a$ yards from it, would be inclined at an angle $a$ to the

horizon. Shew that, if the rifle be wrongly sighted so that the elevation is $(a + \theta)$, where $\theta$ is small compared with $a$, the target will be hit at a height $\dfrac{a \cos 2a}{\cos^2 a} \theta$ above the centre of the bull's eye.

If the range be 960 yards, the time of flight 2 seconds, and the error of elevation $1''$, the height above the centre of the bull's eye at which the target will be hit is nearly $\frac{1}{8}$th of an inch.

21. If a rifle be sighted so as to throw a ball at a small elevation $\delta$ above the line of aim and thus hit an object distant $R$ on a horizontal plane, shew that, in order that the particle may have the same range when firing up or down a given slope whose inclination is $i$, the sights must be altered so as to give an elevation $\delta \cos i$ approximately.

22. If $a$ be the angle of projection in order that a bullet projected with velocity $V$ from a platform at rest may strike an object in the same horizontal plane, shew that when the platform is moving towards the object with a velocity $u$ (which is small compared with $V$) the angle of projection must be diminished by $\dfrac{u \sin a}{V \cos 2a} \dfrac{180°}{\pi}$ nearly, provided that the object is well within range for the given velocity.

23. From a fort a buoy was observed at a depression $i$ below the horizon; a gun was fired at it at an elevation $a$ but the shot was observed to strike the water at a point whose depression was $i'$. Shew that to strike the buoy the gun must be fired at an elevation $\theta$ where

$$\frac{\cos \theta \sin (\theta + i)}{\cos a \sin (a + i')} = \frac{\cos^2 i \sin i'}{\cos^2 i' \sin i}.$$

24. Shew that the locus of the foci of all trajectories which pass through two given points is a hyperbola.

25. Particles are projected from the same point in different directions in the same vertical plane so that (1) the horizontal component of their velocities, (2) the time of flight, is constant. Shew that the locus of the focus is, in each case, a parabola.

26. A heavy particle slides down a chord of a vertical circle to the lowest point and then flies off and describes a parabola. Shew that the focus of the parabola lies in the tangent to the circle at the upper extremity of the chord.

27. Particles descend under gravity along various straight lines from a fixed point to a fixed vertical straight line and then freely describe parabolic trajectories; shew that the locus of the foci of the paths described is a circle and the locus of their vertices an ellipse.

28. A gun on a smooth horizontal plane throws a ball whose mass is $\frac{1}{n}$th of its own mass; shew that, if the momentum of the ball on first leaving the gun be constant, the elevation of the gun giving the greatest range on the horizontal plane through the muzzle is $\tan^{-1}\frac{n}{1+n}$.

29. A shot is fired with velocity $\sqrt{2gh}$ from the top of a mountain which is in the form of a hemisphere of radius $r$. Shew that the furthest points of the mountain which can be reached by the shot are at a distance (measured in a straight line), $r - \sqrt{r^2 - 4rh}$ from the point of projection.

30. With given velocity of projection give a geometrical construction to obtain the maximum range of a ship which can be struck from a fort on the top of a cliff, and the maximum range at which the ship can strike the fort; determine also the angle of projection in each case.

EXAMPLES ON CHAPTER V.   197

31. Shew that the whole area commanded by a gun planted on a hill side, supposed plane, is an ellipse whose focus is at the gun, eccentricity the sine of the inclination of the hill, and semi-latus-rectum equal to twice the height to which the velocity of the shot at the muzzle is due.

32. **A gun** is placed on a fort situated on the side of a hill, which is **a** plane of inclination $a$. Shew that the area commanded by it is $4\pi h \, (h + d \cos a) \sec^3 a$, where $\sqrt{2gh}$ is the muzzle velocity of the shot **and** $d$ **the** perpendicular distance of the gun from the hill side.

33. Shew that the area **covered** on a vertical wall at a distance $a$ by a jet of a fire-engine, placed on level **ground**, is, for all directions **of the jet**, a parabola whose height is $\dfrac{b^2 - a^2}{2b}$ and **breadth** $2\sqrt{b^2 - a^2}$, where $b$ denotes the maximum range of the jet.

34. Find that point in a given horizontal plane from which **a** particle can be projected with the least velocity so that it passes through two given points not in the **horizontal** plane.

35. Shew that, in order to project a particle with **given** velocity $v$ so that it may give the maximum impulse **to a** smooth vertical wall at a height $h$ above the point of projection, the latter must be at a distance $\sqrt{\dfrac{2h}{g}(v^2 - 2gh)}$ from the vertical wall.

36. The ridge of a roof is 15 feet vertically above the eaves and the slope is $45°$; find the greatest vertical depth **below** the eaves at which a gutter projecting six inches beyond **the eaves may be** placed so as to catch all the rain water

coming from the roof, assuming that each drop starts from rest at the point where it strikes the roof, and neglecting all friction.

37. A man standing on the edge of a cliff throws a stone with given velocity $u$ at a given inclination to the horizon, in a plane perpendicular to the edge of the cliff; after an interval $\tau$ he throws from the same spot another stone with given velocity $v$ at an angle $\frac{\pi}{2} + \theta$ with the line of discharge of the first stone and in the same plane. Find $\tau$ so that the stones may strike one another, and shew that the maximum value of $\tau$ for different values of $\theta$ is $2v^2/gw$ and occurs when $\theta = \sin^{-1} v/u$, where $w$ is $v$'s vertical component.

38. A cannon, whose elevation is $a$, points at a target; the trunnions of the gun are then tilted through an angle $\beta$ in the vertical plane of the axis. Shew that, if $R$ be the range, the corresponding horizontal deviation is $R \tan a \sin \beta$.

39. Find the charge of powder required with an elevation of 15° to send a 32 lb. shot to a range of 1600 yards, it being known that the initial velocity of the shot is 1600 feet per second when the charge is half the weight of the shot.

If the gun be moveable on a smooth horizontal stand, and if the weight of the gun be $n$ times the weight of the shot, and the charge that just found, then the range is $\dfrac{6400\,n}{4n + 2 - \sqrt{3}}$ yards.

40. Two parallel lines in the same vertical plane are each inclined to the horizon at an angle $a$. From a point midway between them a particle is projected so as to graze one line and fall perpendicularly on the other; shew that the angle that the direction of projection makes with either is

$$\tan^{-1}\left[(\sqrt{2} - 1) \cot a\right].$$

41. A particle is projected from the lowest point $O$ of a hollow sphere in such a direction and with such a velocity that it strikes the sphere at right angles at some point $P$. If $a$, $\theta$ be the angles which the direction of projection and the line $OP$ respectively make with the horizontal, then

$$2 \tan a = \cot \theta + 3 \tan \theta.$$

42. Two particles of unequal mass are projected so as to describe the same path in opposite directions and when they meet they coalesce; find the latus-rectum and focus of the new parabola that they will describe.

43. A bullet is fired in a direction towards a second equal bullet which is let fall at the same instant. Shew that they meet and that, if they coalesce, the latus-rectum of the joint path is one quarter the latus-rectum of the original path of the first bullet.

44. A thin board, of thickness $a$, is placed in front of a gun and opposes a constant resistance $nmg$ to the motion of a bullet, of mass $m$, during its passage through the board. The barrel of the gun makes an angle $\theta$ with the horizon, and the bullets are fired with a velocity due to a fall through a vertical height $h$; when the board is withdrawn the bullets hit the top of a tower, and when it is present they hit the bottom of the tower. Shew that the height of the tower is $\dfrac{4na}{h} \tan^2 \theta (h \cos \theta - na)$, where $n$ is very large and $na$ finite.

45. A snowball is projected with velocity $V$ horizontally from the edge of a cliff, $a$ feet in height, and passes through a layer of cloud, $b$ feet in thickness, which is moving horizontally with velocity parallel and equal to the initial velocity of the ball; shew that if the mass of the snowball during its passage through the cloud increase in the time $\tau$ in the ratio $1 + \dfrac{g\tau}{u} : 1$,

where $u$ is the vertical velocity of the ball when it reaches the cloud, the range on a horizontal plane passing through the foot of the cliff will be

$$\frac{by + u\sqrt{2g(a-b)}}{ug} V.$$

46. The elevation of a rifle for a parabolic trajectory of range $a$ is $\alpha$; shew that, taking into account a uniform horizontal resistance equal to $\lambda V^2$, where $V$ is the initial velocity, the elevation would have to be increased by

$$\tfrac{1}{3} a \lambda \tan \alpha \sec 2\alpha$$

approximately.

47. A particle is projected from a platform with velocity $V$ at an elevation $\beta$. On it is a telescope fixed at an elevation $\alpha$. The platform moves horizontally in the plane of motion of the particle so as to keep the particle always in the centre of the field of view of the telescope. Shew that the original velocity of the platform must be $V \dfrac{\sin(\alpha - \beta)}{\sin \alpha}$ and its acceleration $g \cot \alpha$.

48. A telescope on a heavy platform is drawn up a smooth plane of inclination $\alpha$ to the horizon by a force so adjusted that the telescope may keep a given projectile always in the field of view. Shew that, if $\beta$ be the angle the telescope makes with the plane and $V$ the initial velocity of the projectile perpendicular to the axis of the telescope, the magnitude of the force per unit mass must be $g \cos \alpha \cot \beta$ and the initial velocity of the telescope $V \cosec \beta$. The motion of all the bodies is supposed to be in one plane.

49. A body, of mass $m$, is projected so as to describe a parabola under the action of gravity. A blow, whose impulse is $mu$, is given to it at any point of its path; shew

that the focus of the new trajectory moves towards the body through a distance $\dfrac{2v-u}{2g}u$, where $v$ is the original velocity of the body.

50. **Two** equal particles are connected by a fine inelastic thread; one is placed on a smooth table and the other just over its edge, the thread being at full stretch at right angles to the edge of the table; find the velocity of the centre of inertia of the particles at the instant after the former has left the table, and shew that the whole interval that elapses from the beginning of the motion to the instant when the thread first becomes horizontal varies as the square root of the length of the thread.

# CHAPTER VI.

### COLLISION OF ELASTIC BODIES.

139. If a man allow a glass ball to drop from his hand upon a marble floor it rebounds to a considerable height, almost as high as his hand; if the same ball be allowed to fall upon a wooden floor it rebounds through a much smaller distance.

If we allow an ivory billiard ball and a glass ball to drop from the same height the distances through which they rebound will be different.

If again we drop a leaden ball upon the same floors the distances through which it rebounds are much smaller than in either of the former cases.

Now the velocities of these bodies are the same on first touching the floor; but, since they rebound through different heights, their velocities on leaving the floor must be different.

The property of the bodies which causes these differences in their velocities after leaving the floor is called their **Elasticity**.

In the present chapter we shall consider some simple cases of the impact of elastic bodies. We can only discuss the cases of particles in collision with particles or planes and of smooth homogeneous spheres in collision with smooth planes or smooth spheres.

**140. Def.** Two bodies are said to *impinge directly* when the direction of motion of each is along the common normal at the point at which they touch.

They are said to *impinge obliquely* when the direction of motion of either or both is not along the common normal at the point of contact.

The direction of this common normal is called the *line of impact*.

In the case of two spheres the common normal is the line joining their centres.

**141. Newton's Experimental Law.** Newton found, by experiment, that if two bodies impinge directly their relative velocity after impact is in a constant ratio to their relative velocity before impact and is in the opposite direction.

If the bodies impinge obliquely their relative velocity resolved along their common normal after impact is in a constant ratio to their relative velocity before impact resolved in the same direction and is of opposite sign.

This constant ratio depends on the substances of which the bodies are made and is independent of the masses of the bodies. It is generally denoted by $e$ and is called the **Modulus** or Coefficient of Elasticity, Restitution, or Resilience. Either of the two latter terms is better than the first, which is used in Physics for other meanings.

If $u$, $u'$ be the component velocities of two bodies before impact along their common normal, and $v$, $v'$ the component velocities of the bodies in the same direction after impact, the law states that $v - v' = -e(u - u')$.

The value of $e$ has widely different values for different bodies; for two glass balls $e$ is ·94; for two ivory ones it is ·81; for two of cork it is ·65; for two of cast-iron about ·66; whilst for two balls of lead it is about ·20 and for two balls, one of lead and the other of iron, the value is ·13.

Bodies for which the coefficient of restitution is zero are said to be "inelastic"; whilst "perfectly elastic" bodies are those for which the coefficient is unity. Probably there are no bodies in nature coming strictly under either of these headings; approximate examples of the former class are such bodies as putty, whilst probably the nearest approach to the latter class is made by glass balls.

More careful experiments have shewn that the ratio of the relative velocities before and after impact is not absolutely constant but decreases very slightly for very large velocities of approach of the bodies.

**142.** *Motion of two smooth bodies perpendicular to the line of impact.*

When two smooth bodies impinge there is no tangential action between them so that the stress between them is entirely along their common normal. Hence there is no force perpendicular to this common normal and therefore no change of velocity in that direction. Hence the component velocity of each body perpendicular to the common normal is unaltered by the impact.

**143.** *Motion of two bodies along the line of impact.*

From Art. 87 it follows that when two bodies impinge

the sum of their momenta along the line of impact is the same after impact as before.

The two principles enunciated in this and the previous article, together **with** Newton's experimental law, are sufficient to find the change in the motion of particles and spheres produced by a collision.

We shall now proceed to the discussion of particular cases.

**144. Impact on a fixed plane.** *A smooth sphere, or particle, whose* **mass is** m *and whose coefficient of restitution is* c, *impinges* **obliquely** *on a fixed plane; to find the change in its motion.*

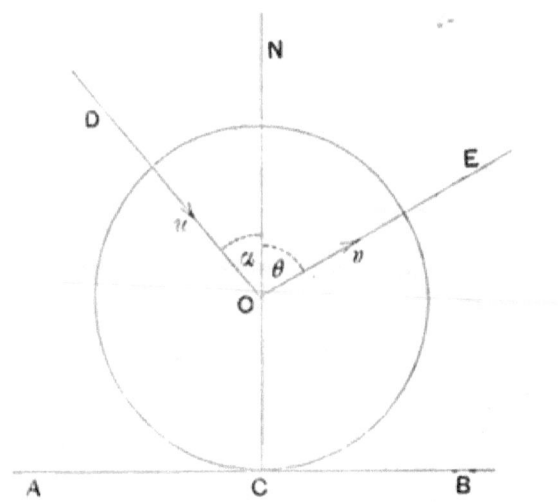

Let $AB$ be the fixed **plane**, $C$ the point at which the sphere impinges and $CN$ the normal to the plane at $C$ so that $CN$ passes through the centre, $O$, of the sphere.

Let $DO$, $OE$ be the directions of motion of the centre of the sphere before and after impact and let the angles

$NOD$, $NOE$ be $\alpha$ and $\theta$. Let $u, v$ be the velocities of the sphere before and after impact as indicated in the figure.

Since the plane is smooth there is no force parallel to the plane; hence the velocity of the sphere resolved in a direction parallel to the plane is unaltered.

$$\therefore v \sin \theta = u \sin \alpha \dots\dots\dots\dots\dots(1).$$

By Newton's experimental law the relative velocity along the common normal after impact is $(-e)$ times the relative velocity before impact.

Hence $\quad v \cos \theta - 0 = -e(-u \cos \alpha - 0)$.

$$\therefore v \cos \theta = eu \cos \alpha \dots\dots\dots\dots\dots(2).$$

From (1) and (2), by squaring and adding, we have

$$v = u \sqrt{\sin^2 \alpha + e^2 \cos^2 \alpha},$$

and, by division, $\cot \theta = e \cot \alpha$.

These two equations give the velocity and direction of motion after impact.

The impulse of the pressure on the plane is equal and opposite to the impulse of the pressure on the sphere and is therefore measured by the change of the momentum of the sphere perpendicular to the plane.

Hence the impulse of the blow $= mu \cos \alpha + mv \cos \theta$

$$= m(1+e) u \cos \alpha.$$

*Cor.* 1. If the impact be direct, we have $\alpha = 0$.

$$\therefore \theta = 0, \text{ and } v = eu.$$

Hence the direction of motion of the sphere is reversed and its velocity reduced in the ratio $1 : e$.

*Cor.* 2. If the coefficient of restitution be unity we have $\theta = \alpha$, and $v = u$.

Hence when the plane is perfectly elastic the angle of reflexion is equal to that of incidence, and the velocity is unaltered in magnitude.

*Cor.* 3. If the coefficient of restitution be zero we have $\theta = 0$, and $v = u \sin \alpha$.

Hence the sphere after impact with an inelastic plane slides along the plane with its velocity parallel to the plane unaltered.

*Cor.* 4. If a smooth sphere impinge on a fixed curved surface the investigation of the impact is reduced to that of this article, the tangent plane to the surface at the point of contact being substituted for the curved surface.

145. If a plane be compelled to move with velocity $V$ and overtake a particle moving with velocity $u$ in the same direction as the plane, the velocity $v$ of the particle after the impact is determined by Newton's law. For, by Newton's law, $V - v = - e(V - u)$.

$$\therefore v = (1 + e) V - eu.$$

If the particle overtake the plane, in which case $u$ is greater than $V$, the same formula gives the velocity of the particle after the impact. In this case, if $V$ be less than $\dfrac{e}{1+e} u$, the velocity of the particle is reversed in direction.

The impulse of the force exerted during the collision

$$= m(v - u) = m(1 + e)(V - u).$$

146. **Direct impact of two spheres.** *A smooth sphere, of mass* m, *impinges directly with velocity* u *on another smooth sphere, of mass* m′, *moving in the same*

*direction with velocity* $u'$. *If the coefficient of restitution be* e, *to find their velocities after the impact.*

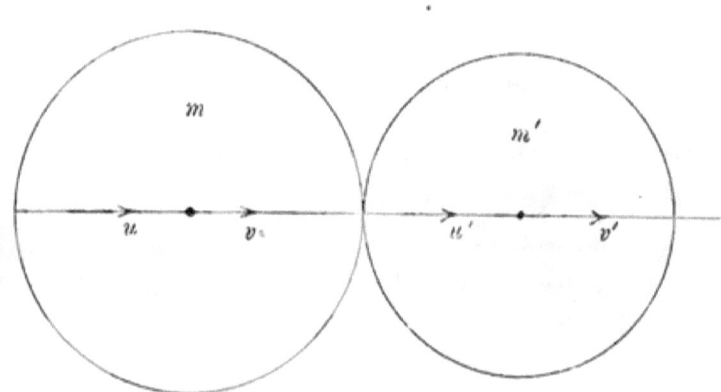

Let $v$, $v'$ be the velocities of the two spheres after impact.

By Newton's experimental law the relative velocity after impact is $(-e)$ times the relative velocity before impact.

$$\therefore v - v' = -e(u - u') \quad \ldots\ldots\ldots\ldots\ldots\ldots(1).$$

Again the only force acting on the bodies during the impact is the blow along the line of centres. Hence, by Art. 87, the total momentum in that direction is unaltered.

$$\therefore mv + m'v' = mu + m'u' \quad \ldots\ldots\ldots\ldots\ldots(2).$$

Multiplying (1) by $m'$, and adding to (2), we have

$$(m + m')v = (m - em')u + m'(1 + e)u'.$$

Again multiplying (1) by $m$, and subtracting from (2), we have

$$(m + m')v' = m(1 + e)u + (m' - em)u'.$$

These two equations give the velocities after impact.

Also the impulse of the blow on the ball $m$

= the change produced in its momentum

$$= m(u-v) = \frac{mm'}{m+m'}(1+e)(u-u').$$

The impulse of the blow on the other ball is equal and opposite to this.

If the second sphere be moving in a direction opposite to that of the first we must change the sign of $u'$.

*Cor.* If we put $m = m'$ and $e = 1$, we have

$$v = u', \text{ and } v' = u.$$

Hence if two equal perfectly elastic balls impinge directly they interchange their velocities.

**147. Oblique impact of two spheres.** *A smooth sphere, of mass* m, *impinges with a velocity* u *obliquely on a smooth sphere, of mass* m′, *moving with velocity* u′. *If the directions of motion before impact make angles* α, β *respectively with the line joining the centres of the spheres and if the coefficient of restitution be* e, *to find the velocities and directions of motion after impact.*

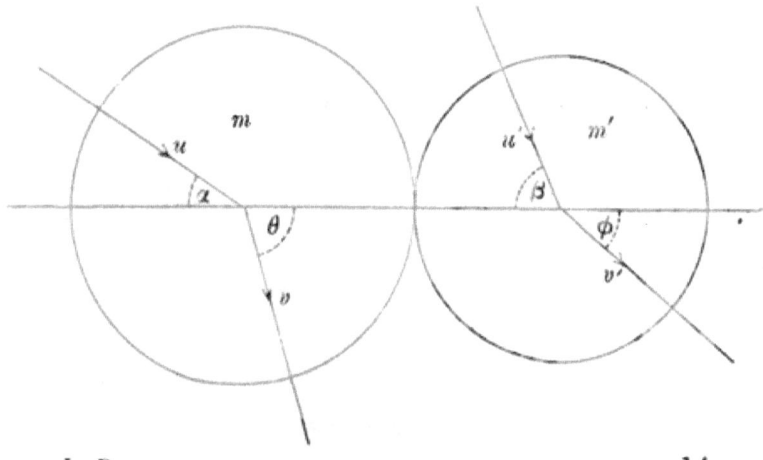

L. D.    14

## OBLIQUE IMPACT OF SPHERES.

Let the velocities of the spheres after impact be $v$, $v'$ in directions inclined at angles $\theta$, $\phi$ to the line of centres.

Since the spheres are smooth there is no force perpendicular to the line joining the centres of the two balls, and therefore the velocities in that direction are unaltered.

Hence
$$v \sin \theta = u \sin \alpha \quad \ldots\ldots\ldots\ldots\ldots\ldots(1),$$
and
$$v' \sin \phi = u' \sin \beta \ldots\ldots\ldots\ldots\ldots\ldots(2).$$

For the motion along the line of centres we have by Newton's Law,
$$v \cos \theta - v' \cos \phi = -e(u \cos \alpha - u' \cos \beta)\ldots\ldots(3).$$

Again, the only force acting on the spheres during the impact is the blow along the line of centres. Hence (Art. 87) the total momentum in that direction is unaltered.

$$\therefore mv \cos \theta + m'v' \cos \phi = mu \cos \alpha + m'u' \cos \beta \ldots\ldots(4).$$

The equations (1), (2), (3), (4) determine the unknown quantities $v$, $v'$, $\theta$, $\phi$.

Multiply (3) by $m'$, add to (4), and we obtain
$$v \cos \theta = \frac{(m - em')u \cos \alpha + m'(1+e)u' \cos \beta}{m + m'} \ldots\ldots(5).$$

So multiplying (3) by $m$, and subtracting from (4), we get
$$v' \cos \phi = \frac{m(1+e)u \cos \alpha + (m' - em)u' \cos \beta}{m + m'} \ldots\ldots(6).$$

From (1), (5) by squaring and adding we obtain $v^2$, and by division we have $\tan \theta$.

Similarly from (2), (6) we obtain $v'^2$ and $\tan \phi$.

Hence the motion is completely determined.

The impulse of the blow on the first ball = the change produced in its momentum $= m (u \cos \alpha - v \cos \theta)$

$$= \frac{mm'}{m+m'} (1+e) (u \cos \alpha - u' \cos \beta), \text{ on reduction.}$$

The impulse of the blow on the other ball is equal and opposite to this.

*Cor.* 1. If $u' = 0$ we have from equation (2) $\phi = 0$, and hence the sphere $m'$ moves along the line of centres. This follows independently, since the only force on $m'$ is along the line of centres.

*Cor.* 2. If $m = m'$ and $e = 1$, we have

$$v \cos \theta = u' \cos \beta, \text{ and } v' \cos \phi = u \cos \alpha.$$

Hence if two equal perfectly elastic spheres impinge they interchange their velocities in the direction of the line of centres.

148. The case of the impact of a particle on a smooth fixed plane may be deduced from that of two smooth spheres.

If in the last article we make $m'$ infinite and $u'$ zero, we have the case of a fixed plane. For a sphere of infinite mass at rest may be looked upon as immovable, and therefore, in the limit, as a fixed plane.

In this case the equations (2) and (6) vanish and the equations (1) and (5) become $v \sin \theta = u \sin \alpha$,

and
$$v \cos \theta = Lt_{m' = \infty} \frac{(m - em') u \cos \alpha}{m + m'}$$

$$= Lt_{m' = \infty} \frac{\left(\frac{m}{m'} - e\right) u \cos \alpha}{\frac{m}{m'} + 1}$$

$$= -eu \cos \alpha.$$

These two equations are now the same as those of Art. 144 if we remember that the velocity of the particle after impact is here measured in the direction opposite to that in Art. 144.

## EXAMPLES.

**1.** *Two smooth balls, one of mass double that of the other, are moving with equal velocities in opposite parallel directions and impinge, their directions of motion at the instant of impact making angles of 30° with the line of centres. If the coefficient of restitution be $\frac{1}{2}$, find the velocities and directions of motion after the impact.*

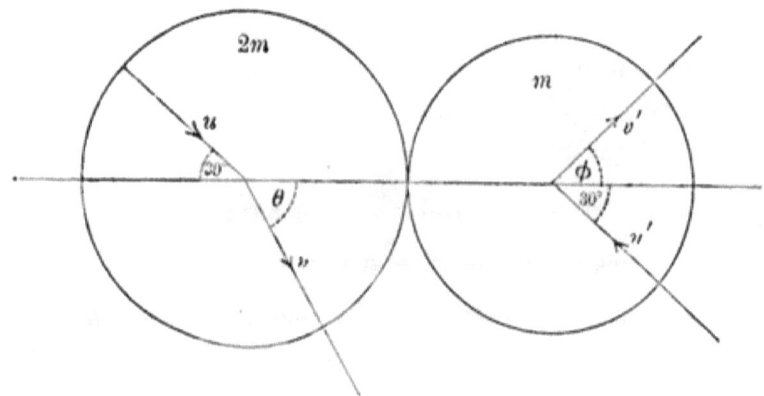

Let the masses of the balls be $2m$, $m$, and let the velocities after impact be $v$, $v'$ respectively at angles $\theta$, $\phi$ to the line of centres.

Since the velocities perpendicular to the line of centres are unaltered,

$$\therefore\ v \sin \theta = u \sin 30° = \frac{u}{2} \quad \ldots\ldots\ldots\ldots\ldots\ldots (1),$$

and
$$v' \sin \phi = u \sin 30° = \frac{u}{2}. \quad \ldots\ldots\ldots\ldots\ldots\ldots (2).$$

By Newton's law,

$$v \cos \theta - v' \cos \phi = -e\,[u \cos 30° - (-u \cos 30°)] = -u\,\frac{\sqrt{3}}{2}\ldots\ldots (3).$$

Since the momentum resolved parallel to the line of centres remains unaltered,

$$\therefore\ 2mv \cos \theta + mv' \cos \phi = 2mu \cos 30° - mu \cos 30°,$$

$$\therefore\ 2v \cos \theta + v' \cos \phi = u\,\frac{\sqrt{3}}{2}\ldots\ldots\ldots\ldots\ldots\ldots (4).$$

Solving (3) and (4) we have $v \cos \theta = 0$, $v' \cos \phi = u\,\frac{\sqrt{3}}{2}$.

# EXAMPLES. 213

From these equations and (1) (2) we obtain

$$\theta = \frac{\pi}{2}, \ v = \frac{u}{2}; \ \phi = 30°, \ v' = u.$$

Hence after impact the larger ball starts off in a direction perpendicular to the line of centres with half its former velocity, and the smaller ball moves as if it were a perfectly elastic ball impinging on a fixed plane.

Also the impulse of the blow

= change in the momentum along the line of centres

$= 2mu \cos 30° = mu\sqrt{3}.$

2. *An imperfectly elastic particle is projected from the foot of a plane, which is inclined to the horizon at an angle $\beta$, with a velocity* u *at an angle* a *to the plane. If* e *be the coefficient of restitution, find the time that elapses before the particle has ceased rebounding and the distance described by it parallel to the plane in this time.*

*Find also the condition that it may return to the point of projection after rebounding* n *times.*

The velocity of the particle perpendicular to the plane initially is $u \sin a$, and this also is its velocity in the opposite direction just before the first impact. Hence (Art. 144) its velocity perpendicular to the plane immediately after the first impact is $eu \sin a$; after the second impact it is $e^2 u \sin a$, and so for the velocities after the other impacts. These velocities form a descending geometrical progression.

Also the acceleration perpendicular to the inclined plane is $-g \cos \beta$.

Hence the times of describing its trajectories are

$$\frac{2u \sin a}{g \cos \beta}, \ \frac{2eu \sin a}{g \cos \beta}, \ \frac{2e^2 u \sin a}{g \cos \beta}, \ \ldots\ldots$$

The sum of these times

$$= \frac{2u \sin a}{g \cos \beta} (1 + e + e^2 + \ldots\ldots \text{ ad inf.}) = \frac{2u \sin a}{g \cos \beta} \cdot \frac{1}{1-e}.$$

After this time the particle ceases to rebound.

Also the distance described parallel to the plane in this time

$$= u \cos a \cdot \frac{2u \sin a}{g \cos \beta} \frac{1}{1-e} - \frac{1}{2} g \sin \beta \cdot \left( \frac{2u \sin a}{g \cos \beta} \frac{1}{1-e} \right)^2$$

$$= \frac{2u^2 \sin a}{g \cos^2 \beta (1-e)^2} [\cos (a + \beta) - e \cos a \cos \beta].$$

Again the time, $\tau$, of describing $n+1$ trajectories

$$= \frac{2u \sin \alpha}{g \cos \beta}[1+e+e^2+\ldots\ldots+e^n] = \frac{2u \sin \alpha}{g \cos \beta} \cdot \frac{1-e^{n+1}}{1-e}.$$

If the particle return to the point of projection after rebounding $n$ times the distance described parallel to the plane in time $\tau$ must be zero.

$$\therefore 0 = u \cos \alpha \tau - \tfrac{1}{2} g \sin \beta \tau^2,$$

$$\therefore \frac{2u \cos \alpha}{g \sin \beta} = \tau = \frac{2u \sin \alpha}{g \cos \beta} \cdot \frac{1-e^{n+1}}{1-e}.$$

$$\therefore \tan \alpha \tan \beta (1 - e^{n+1}) = 1 - e.$$

3. If a ball overtake a ball of twice its own mass moving with one-seventh of its velocity and if the coefficient of restitution between them be $\tfrac{3}{4}$, the first ball will, after striking the second ball, remain at rest.

4. A ball of mass 2 lbs. impinges directly on a ball of mass 1 lb. which is at rest; find the coefficient of restitution if the velocity with which the larger ball impinges be equal to the velocity of the smaller ball after impact.

*Ans.* $\tfrac{1}{2}$.

5. A ball of mass $m$ impinges directly upon a ball of mass $m_1$ at rest; the velocity of $m$ after impact is $\tfrac{2}{5}$ths of its velocity before impact and the modulus of elasticity is $\tfrac{2}{3}$; compare (i) the masses of the two balls, and (ii) the velocities of $m$ and $m_1$ after impact.

*Ans.* (i) They are as $3:1$; (ii) they are as $1:2$.

6. If the masses of two balls be as $2:1$, and their respective velocities before impact be as $1:2$ and in opposite directions, and $e$ be $\tfrac{5}{6}$, shew that each ball will after direct impact move back with $\tfrac{5}{6}$ths of its original velocity.

7. Two equal marbles, $A$ and $B$, lie in a smooth horizontal circular groove at opposite ends of a diameter; $A$ is projected along the groove and at the end of time $t$ impinges on $B$; shew that a second impact will occur at the end of time $\dfrac{2t}{e}$.

8. A heavy elastic ball drops from the ceiling of a room and after twice rebounding from the floor reaches a height equal to one-half that of the room; shew that its coefficient of restitution is $\sqrt[4]{\tfrac{1}{2}}$.

# EXAMPLES.

9. A particle falls from a height $h$ upon a fixed horizontal plane; if $e$ be the coefficient of restitution shew that the whole distance described by the particle before it has finished rebounding is $\frac{1+e^2}{1-e^2} h$, and that the time that elapses is $\sqrt{\frac{2h}{g}} \frac{1+e}{1-e}$.

10. The masses of five balls at rest in a straight line form a geometrical progression whose ratio is 2 and their coefficients of restitution are each $\frac{2}{3}$. If the first ball be started towards the second, shew that the velocity communicated to the fifth is $(\frac{8}{9})^4 u$.

11. A series of $n$ elastic spheres whose masses are proportional to 1, $e$, $e^2$, ...... are lying in a straight line on a smooth table; if the first impinge directly on the second with velocity $u$, find the resulting velocity of the last, and shew that the final kinetic energy is $\frac{m}{2} u^2 (1-e+e^n)$ where $m$ is the mass of the first ball.

12. A ball of given elasticity slides from rest down a smooth inclined plane, of length $l$, which is inclined at an angle $\alpha$ to the horizon and impinges on a fixed smooth horizontal plane at the foot of the former; find its range on the horizontal plane.

*Ans.* $4el \cos \alpha \sin^2 \alpha$.

13. A heavy elastic ball falls from a height of $n$ feet and meets a plane inclined at an angle of 60° to the horizon; find the distance between the first two points at which it strikes the plane.

*Ans.* $2\sqrt{3} ne (1+e)$ feet.

14. A particle is projected along a smooth horizontal plane from a given point $A$ in it, so that after impinging on an imperfectly elastic vertical plane it may pass through another given point $B$ of the horizontal plane; give a geometrical construction for the direction of projection.

*Ans.* Draw $BN$ perpendicular to the vertical plane and produce to $C$ so that $BN = e \cdot CN$; the required direction of projection is $AC$.

15. A sphere of mass $m$ impinges obliquely on a sphere of mass $M$. Shew that, if $m = eM$, the directions of motion of the spheres after impact are at right angles.

16. A sphere impinges on a sphere of equal mass which is at rest; if the directions of motion after impact be inclined at angles of 30° to the original direction of motion of the impinging sphere, shew that the coefficient of restitution is $\frac{1}{3}$.

**17.** A ball impinges on another equal ball moving with the same speed in a direction perpendicular to its own, the line joining the centres of the balls at the instant of impact being perpendicular to the direction of motion of the second ball; if $e$ be the coefficient of restitution, shew that the direction of motion of the second ball is turned through an angle $\tan^{-1}\dfrac{1+e}{2}$.

**18.** If two imperfectly elastic spheres impinge on one another, find the condition that the direction and velocity of one ball after impact may be the same as the direction and velocity of the other before impact.

*Ans.* The velocities of the spheres before impact perpendicular to the line of centres must be the same and their masses in the ratio $e:1$, where $e$ is the coefficient of restitution.

**19.** Two equal smooth elastic spheres moving in opposite parallel directions impinge on one another; if the inclination of their directions of motion to the line of centres be $\tan^{-1}\sqrt{e}$, where $e$ is the coefficient of restitution, shew that their directions of motion will be turned through a right angle.

**20.** Two equal balls are in contact on a table; a third equal ball strikes them simultaneously and remains at rest after the impact; shew that the coefficient of restitution is $\frac{2}{3}$.

**21.** Two elastic spheres, $P$ and $Q$, of masses 1 lb. and 2 lbs. respectively, are placed on a smooth horizontal plane. $P$ impinging directly on $Q$ at rest drives it perpendicularly against a hard vertical wall from which it rebounds and meets $P$ when the interval between $Q$ and the wall is one-half what it originally was; find the modulus of elasticity which is the same for the impact between the spheres and the plane.

*Ans.* $\frac{1}{10}(\sqrt{21}-1)$.

**22.** An imperfectly elastic particle is projected from a point in a horizontal plane with velocity $u$ at an elevation $a$; if $e$ be the coefficient of restitution, shew that it ceases to rebound from the plane at the end of time $\dfrac{2u}{g}\dfrac{\sin a}{1-e}$.

**23.** A sphere, $A$, impinges directly on a sphere, $B$, and $B$ then impinges directly on a third sphere, $C$; if the spheres be imperfectly elastic, shew that the velocity communicated to $C$ is a maximum when the mass of $B$ is a mean proportional between the masses of $A$ and $C$.

**149. Loss of Kinetic Energy by Impact.** *Two spheres of given masses moving with given velocities impinge; to shew that there is a loss of kinetic energy and to find the amount.*

I. Let the collision be direct and the notation as in Art. 146.

Then we have

$$mv + m'v' = mu + m'u' \quad \ldots \ldots \ldots \ldots \ldots (1),$$

$$v - v' = -e(u - u') \quad \ldots \ldots \ldots \ldots \ldots (2).$$

To the square of (1) add the square of (2) multiplied by $mm'$; we then have

$$(m^2 + mm')v^2 + (m'^2 + mm')v'^2$$
$$= (mu + m'u')^2 + e^2 mm'(u - u')^2,$$

or $\quad (m + m')(mv^2 + m'v'^2)$
$$= (mu + m'u')^2 + mm'(u - u')^2 - (1 - e^2) mm'(u - u')^2$$
$$= (m + m')(mu^2 + m'u'^2) - (1 - e^2) mm'(u - u')^2.$$

$\therefore \tfrac{1}{2}mv^2 + \tfrac{1}{2}m'v'^2 = \tfrac{1}{2}mu^2 + \tfrac{1}{2}m'u'^2 - \dfrac{1-e^2}{2}\dfrac{mm'}{m+m'}(u-u')^2.$

$\therefore$ kinetic energy after impact = kinetic energy before impact $-\dfrac{1-e^2}{2}\dfrac{mm'}{m+m'}(u-u')^2.$

Hence the loss of kinetic energy is

$$\dfrac{1-e^2}{2}\dfrac{mm'}{m+m'}(u-u')^2,$$

and this loss does not vanish unless $e = 1$, that is, unless the balls are perfectly elastic.

## 218 LOSS OF KINETIC ENERGY.

II. Let the impact be oblique and the notation as in Art. 147.

Then we have  $v \sin \theta = u \sin \alpha$ ......................(1),

$$v' \sin \phi = u' \sin \beta ......................(2),$$

$$mv \cos \theta + m'v' \cos \phi = mu \cos \alpha + m'u' \cos \beta ...(3),$$

$$v \cos \theta - v' \cos \phi = - e (u \cos \alpha - u' \cos \beta)...(4).$$

To the square of (3) add the square of (4) multiplied by $mm'$; we then have

$$(m^2 + mm') v^2 \cos^2 \theta + (m'^2 + mm') v'^2 \cos^2 \phi$$
$$= (mu \cos \alpha + m'u' \cos \beta)^2 + e^2 mm' (u \cos \alpha - u' \cos \beta)^2$$
$$= (mu \cos \alpha + m'u' \cos \beta)^2 + mm' (u \cos \alpha - u' \cos \beta)^2$$
$$- (1 - e^2) mm' (u \cos \alpha - u' \cos \beta)^2.$$

$\therefore$, as in I.,

$$\tfrac{1}{2} mv^2 \cos^2 \theta + \tfrac{1}{2} m'v'^2 \cos^2 \phi = \tfrac{1}{2} mu^2 \cos^2 \alpha + \tfrac{1}{2} m'u'^2 \cos^2 \beta$$

$$- \frac{1 - e^2}{2} \frac{mm'}{m + m'} (u \cos \alpha - u' \cos \beta)^2 ............(5).$$

Again, adding together the square of equation (1) multiplied by $\frac{m}{2}$, the square of equation (2) multiplied by $\frac{m'}{2}$, and equation (5), we have

$$\tfrac{1}{2} mv^2 + \tfrac{1}{2} m'v'^2$$
$$= \tfrac{1}{2} mu^2 + \tfrac{1}{2} m'u'^2 - \frac{1 - e^2}{2} \frac{mm'}{m + m'} (u \cos \alpha - u' \cos \beta)^2,$$

or kinetic energy after impact = kinetic energy before impact $- \dfrac{1 - e^2}{2} \dfrac{mm'}{m + m'} (u \cos \alpha - u' \cos \beta)^2$,

and therefore the energy lost is

$$\frac{1 - e^2}{2} \frac{mm'}{m + m'} (u \cos \alpha - u' \cos \beta)^2.$$

Hence in any impact except where the coefficient of restitution is unity we see that energy is lost.

This missing kinetic energy is converted into molecular energy and chiefly reappears in the shape of heat.

**150. Action between two elastic bodies during their collision.** When two elastic bodies impinge the time during which the impact lasts may be divided into two parts, during the first of which the bodies are compressing one another, and during the second of which they are recovering their shape. That the bodies are compressed may be shewn experimentally by dropping a billiard ball upon a floor which has been covered with fine coloured powder. At the spot where the ball hits the floor the powder will be found to be removed not from a geometrical point only but from a small circle; this shews that at some instant during the compression the part of the ball in contact with the floor was a circle; it follows that the ball was then deformed and afterwards recovered its shape.

The first portion of the impact lasts until the bodies are instantaneously moving with the same velocity; forces then come into play tending to make the bodies recover their shape. The mutual action between the bodies during the first portion of the impact is often called "the force of compression," and that during the second portion "the force of restitution."

It is easy to shew that the ratio of the impulses of the forces of restitution and compression is equal to the quantity $e$ which we have defined as the coefficient of restitution.

For consider the case of one sphere impinging directly on another as in Art. 146 and use the same notation.

Let $U$ be the common velocity of the bodies at the instant when the compression is finished. Then

$m(u - U)$ is the loss of momentum by the first ball,

and $m'(U - u')$ is the gain by the second ball.

∴ if $I$ be the impulse of the force of compression we have
$$I = m(u - U) = m'(U - u'),$$
$$\therefore \frac{I}{m} + \frac{I}{m'} = u - U + U - u' = u - u' \ldots\ldots\ldots(1).$$

Again the loss of momentum by the first ball during the period of restitution is $m(U - v)$, and the gain by the second ball is $m'(v' - U)$.

∴ if $I'$ be the impulse of the force of restitution
$$I' = m(U - v) = m'(v' - U),$$
$$\therefore \frac{I'}{m} + \frac{I'}{m'} = U - v + v' - U = v' - v \ldots\ldots\ldots(2).$$

∴ from (1) and (2) $\quad \dfrac{I'}{I} = \dfrac{v' - v}{u - u'} = e.$

$$\therefore I' = eI.$$

**151.** When the bodies which come into collision are rough, rotations are set up in them and the motion is much more complicated than when they are smooth. We cannot discuss the general case within the limits of this book; the only case we shall consider is that of a particle impinging on a rough plane.

**Impact of a particle on a rough plane.** *A particle impinges on a fixed rough plane whose coefficient of friction is $\mu$; to find the resulting change in the motion.*

Let $u, v$ be the velocities before and after impact, the

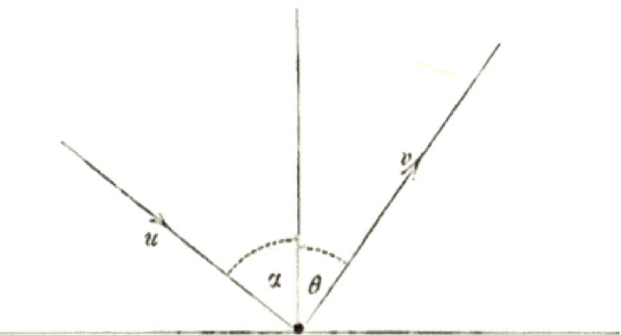

directions of motion being inclined at angles $\alpha, \theta$ to the normal to the plane at the point of impact; let $m$ be the mass of the particle.

At each instant during the time the impact lasts the tangential force is $\mu$ times the normal force.

∴ the total tangential impulse is equal to $\mu$ times the normal impulse.

∴ $m(u \sin \alpha - v \sin \theta) = \mu m (u \cos \alpha + v \cos \theta)$.

∴ $v(\mu \cos \theta + \sin \theta) = u[\sin \alpha - \mu \cos \alpha]$ ........(1).

Also, by Newton's law,

$$v \cos \theta = eu \cos \alpha \quad \text{..................(2)}.$$

From (1), (2) by division

$$\frac{\mu \cos \theta + \sin \theta}{\cos \theta} = \frac{\sin \alpha - \mu \cos \alpha}{e \cos \alpha},$$

or  $e \tan \theta = \tan \alpha - \mu (1 + e)$ ........(3).

Also substituting this value of $\theta$ in (2) we obtain the value of $v$.

From (3) it follows that the greater $\mu$ the smaller is $\theta$ and then by (2) the less is $v$. Hence the rougher the plane the more nearly does the ball rise from it perpendicularly and the less is the velocity of the ball after the collision.

The student might conclude from equation (3) that it might be possible, for some values of $e$, $\mu$, and $\alpha$, that the value of $\theta$ might be negative, and hence that the particle might rebound so that its motion was on the same side of the normal after impact as before the impact. This however is not the case; for friction, being a *passive* force, only acts so as to prevent, or destroy, the velocity of the particle parallel to the plane. Hence if at any instant during the time that the collision lasts the velocity parallel to the plane have been destroyed, the force of friction immediately vanishes; it could not communicate to the particle a velocity parallel to the plane in the direction in which it itself acts.

Ex. 1. A particle impinges on a rough fixed plane, whose coefficient of friction is $\dfrac{1}{\sqrt{3}}$, in a direction making an angle of 60° with the normal to the plane; if the coefficient of restitution be $\frac{1}{2}$, shew that the angle of reflexion of the particle is the same as its angle of incidence and that it loses half its velocity.

Ex. 2. A particle impinges with velocity $u$ on a rough fixed plane whose coefficient of friction is $\frac{2}{3}$, in a direction making an angle of 45° with the normal to the plane; if the coefficient of restitution be $\frac{1}{2}$, shew that the particle rebounds at right angles to the plane with velocity $\dfrac{u}{2\sqrt{2}}$.

## MISCELLANEOUS EXAMPLES.

1. *A smooth ring is fixed horizontally on a smooth table and from a point of the ring a particle is projected along the surface of the table. If $e$ be the coefficient of restitution between the ring and the particle, shew that*

the latter will after three rebounds return to the point of projection if its initial direction of projection make an angle $\tan^{-1} e^{\frac{3}{2}}$ with the normal to the ring.

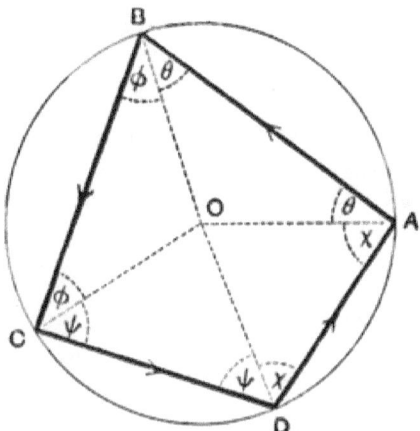

Let $ABCD$ be the path of the particle and $O$ the centre of the circle. Let the angles $OAB$, $OBC$, $OCD$, $ODA$ be $\theta$, $\phi$, $\psi$, $\chi$ respectively. Then $ABO$, $BCO$, $CDO$ are also $\theta$, $\phi$, $\psi$ respectively. If the particle return to $A$ then $DAO$ is $\chi$ also.

We have $\quad \cot\phi = e\cot\theta$; $\cot\psi = e\cot\phi$; $\cot\chi = e\cot\psi$ ......(1);

$$\therefore \tan\phi = \frac{1}{e}\tan\theta;\ \tan\psi = \frac{1}{e^2}\tan\theta;\ \tan\chi = \frac{1}{e^3}\tan\theta.$$

But, since $ABCD$ is a quadrilateral inscribed in a circle,

$$\therefore \theta + \phi + \psi + \chi = \pi.$$

$\therefore \tan\theta + \tan\phi + \tan\psi + \tan\chi = \tan\theta\tan\phi\tan\psi + \tan\theta\tan\phi\tan\chi$
$\qquad\qquad\qquad\qquad + \tan\theta\tan\psi\tan\chi + \tan\phi\tan\psi\tan\chi.$

$$\therefore \tan\theta\left[1 + \frac{1}{e} + \frac{1}{e^2} + \frac{1}{e^3}\right] = \tan^3\theta\left[\frac{1}{e^3} + \frac{1}{e^4} + \frac{1}{e^5} + \frac{1}{e^6}\right]$$

$$= \tan^3\theta \cdot \frac{1}{e^3}\left[1 + \frac{1}{e} + \frac{1}{e^2} + \frac{1}{e^3}\right].$$

$$\therefore \tan\theta = e^{\frac{3}{2}}.$$

It may be noted that, since $\tan\phi\tan\psi = \tan\chi\tan\theta = \frac{1}{e^3}\tan^2\theta = 1$, the angles $BCD$, $DAB$ are right angles.

**2.** *A particle is placed within a straight tube, of length* a *and of small section, and the ends of the tube are closed. If the tube be placed on a smooth horizontal plane and projected in the direction of its length, find the distance advanced by the tube between the first and the* $(n+1)$th *impact, the masses of the tube and particle being equal and the coefficient of restitution between them being* e.

Let $u$ be the velocity with which the tube is projected.

The relative velocities of the tube and shot after the 1st, 2nd,... impacts are respectively $-eu, e^2u, -e^3u$.......

Hence the times that elapse between successive impacts are

$$\frac{a}{eu}, \frac{a}{e^2u}, \frac{a}{e^3u} \dots\dots$$

∴ time between first and $(n+1)$th impact

$$= \frac{a}{u}[e^{-1} + e^{-2} + \dots + e^{-n}] = \frac{a}{eu}\left[\frac{e^{-n}-1}{e^{-1}-1}\right].$$

Also throughout the motion the velocity of the centre of inertia is $\frac{u}{2}$, and hence the distance described by it

$$= \frac{u}{2} \cdot \frac{a}{eu} \cdot \frac{e^{-n}-1}{e^{-1}-1} = \frac{a}{2} \cdot \frac{e^{-n}-1}{1-e}.$$

If $n$ be even, the particle is at the $(n+1)$th impact in contact with the end of the tube which initially struck it; if $n$ be odd, the particle is in contact with the other end of the tube and the centre of inertia has described a distance greater by $\frac{a}{2}$ than the tube has.

Hence the space described by the tube is

$$\frac{a}{2}\frac{e^{-n}-1}{1-e} \text{ or } \frac{a}{2}\frac{e^{-n}-1}{1-e} - \frac{a}{2},$$

according as $n$ is even or odd.

**3.** *A particle is projected from a point in a smooth plane inclined to the horizon at an angle* a, *in a vertical plane cutting the inclined plane in a horizontal line, and in a direction inclined to the horizon at an angle* $\theta$. *Shew that at the nth rebound the distance described by the particle parallel to the line of greatest slope is* a sin a tan $\theta \dfrac{e(1-e^{n-1})}{1-e}$, *where* a *is the horizontal distance described and* e *is the coefficient of restitution.*

If $u$ be the initial velocity of the particle, its initial horizontal velocity is $u \cos \theta$, and its initial vertical velocity is $u \sin \theta$.

The latter is equivalent to two components, $u \sin \theta \cos \alpha$ perpendicular to the inclined plane, and $u \sin \theta \sin \alpha$ parallel to the line of greatest slope.

Also the acceleration in these two directions is $g \cos \alpha$ and $-g \sin \alpha$ respectively.

After each impact the velocity perpendicular to the plane is altered in the ratio of 1 to $e$.

Hence the time, $T$, that elapses before the $n$th impact

$$= \frac{2u \sin \theta \cos \alpha}{g \cos \alpha} [1+e+\ldots+e^{n-1}] = \frac{2u \sin \theta}{g} \frac{1-e^n}{1-e}.$$

During this time the horizontal velocity remains unaltered,

$$\therefore a = u \cos \theta \cdot T.$$

Also if $x$ be the distance described parallel to the line of greatest slope we have $\qquad x = u \sin \theta \sin \alpha \cdot T - \tfrac{1}{2} g \sin \alpha T^2$.

$$\therefore \frac{x}{a} = \sin \alpha \left[ \tan \theta - \frac{g}{2u \cos \theta} \cdot T \right] = \sin \alpha \tan \theta \left[ 1 - \frac{1-e^n}{1-e} \right]$$

$$\therefore x = -a \sin \alpha \tan \theta \frac{e(1-e^{n-1})}{1-e}.$$

4. *A ball is projected from a point* A *at an elevation* $\alpha$ *against a rough vertical wall and in a plane perpendicular to the wall; shew that after impact with the wall it will return to the point of projection if* $ga(1+e) \cos \lambda = 2eV^2 \cos \alpha \sin(\alpha - \lambda)$, *where* V *is the velocity of projection,* $a$ *the distance of* A *from the wall,* $e$ *the coefficient of restitution, and* $\tan \lambda$ *the coefficient of friction between the ball and the wall.*

The time, $T$, in the first trajectory is $\dfrac{a}{V \cos \alpha}$ and the time, $T_1$, in the second is $\dfrac{a}{eV \cos \alpha}$.

Also the impulse of the blow normal to the wall is $(1+e) V \cos \alpha$ and therefore the tangential impulse is $\mu (1+e) V \cos \alpha$.

Hence if $U$ be the vertical velocity after the impact we have

$$\mu (1+e) V \cos \alpha = V \sin \alpha - g T - U \ldots\ldots\ldots\ldots\ldots\ldots(1).$$

Now in the time $T + T_1$ the whole distance described vertically is zero.

$$\therefore 0 = V \sin \alpha T - \tfrac{1}{2} g T^2 + U T_1 - \tfrac{1}{2} g T_1^2,$$

L. D.     15

or by (1), substituting for $U$,

$$0 = V \sin a (T+T_1) - \tfrac{1}{2}g(T+T_1)^2 - \mu(1+e) V \cos a T_1.$$

$$\therefore 0 = V \sin a - \tfrac{1}{2}g \overline{1+e}\, T_1 - \mu V \cos a, \text{ since } T = eT_1.$$

$$\therefore V \sin(a-\lambda) = \frac{g}{2}\overline{1+e}\cos\lambda \,.\, T_1 = \frac{g\,\overline{1+e}\cos\lambda}{2} \cdot \frac{a}{eV \cos a},$$

$$\therefore 2eV^2 \cos a \sin(a-\lambda) = ga(1+e)\cos\lambda.$$

**5.** *An inclined plane, of mass* M, *is capable of moving freely on a smooth horizontal plane and a smooth sphere, of mass* m, *is dropped upon its face and rebounds; shew that the initial velocity of the plane is* m$(1+e)$ u sin a cos a $/$ $(M + $m$\sin^2 a)$, *where* u *is the velocity of the sphere on striking the plane,* a *is the inclination of the face of the plane, and* e *is the coefficient of restitution.*

Immediately after the impact let $U_1$, $U_2$ be the components of the velocity of the ball respectively perpendicular to and along the plane, and let $V$ be the velocity of the plane.

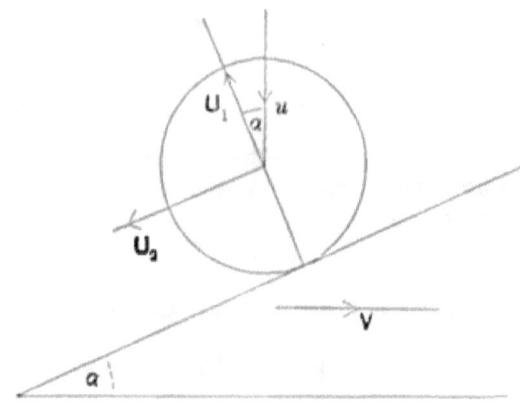

Since the plane is smooth, we have

$$U_2 = u \sin a \quad\quad\quad\quad\quad\quad\quad\quad\quad\quad (1).$$

Also by Newton's Law, we have

$$-U_1 - V \sin a = -eu \cos a \quad\quad\quad\quad\quad\quad (2).$$

Also the impulse perpendicular to the inclined face $= m(u\cos a + U_1)$ and the horizontal component of this impulse is equal to the momentum generated in the plane.

$$\therefore m(u \cos a + U_1)\sin a = MV$$

or
$$MV - mU_1 \sin\alpha = mu \cos\alpha \sin\alpha \quad \ldots\ldots\ldots\ldots\ldots(3).$$

From (2), (3), by solving, we have
$$V(M + m\sin^2\alpha) = mu(1+e)\cos\alpha\sin\alpha \ldots\ldots\ldots(4)$$
and
$$U_1[M + m\sin^2\alpha] = u\cos\alpha\,(eM - m\sin^2\alpha)\ldots\ldots\ldots(5).$$

Equation (4) gives the velocity of the plane and (1), (5) give the velocity and direction of the sphere.

## EXAMPLES. CHAPTER VI.

1. SHEW that, in order to produce the maximum deviation in the direction of a smooth billiard ball of diameter $a$ by impact on another equal ball at rest, the former must be projected in a direction making an angle $\sin^{-1}\dfrac{a}{c}\sqrt{\dfrac{1-e}{3-e}}$ with the line of length $c$ joining the two centres, and prove also that this maximum deviation is $\tan^{-1}\dfrac{1+e}{2\sqrt{2}\sqrt{1-e}}$, where $e$ is the coefficient of restitution.

2. A series of balls, whose masses are $M_1, M_2, \ldots\ldots$, are arranged with their centres in a straight line and the coefficient of restitution between the $r$th and $(r+1)$th is $\dfrac{M_{r+1}}{M_r}$; shew that, if $M_1$ impinge directly on $M_2$ at rest and so on, the velocity of each ball between the impacts is equal to the initial velocity of $M_1$.

3. The red ball is standing in the middle of a billiard table whose length is $2a$ and breadth $2b$; shew that the direction in which the white ball must be sent from baulk so that it may hole the red ball in the top pocket and itself in the other is inclined to the side of the table at an angle $\tan^{-1}\dfrac{b\,(ea^2 - b^2)}{a\,(a^2 - eb^2)}$, and hence find limits to the possibility of the stroke.

4. An inelastic ball, of small radius, sliding along a smooth horizontal plane with a velocity of 16 feet per second impinges on a smooth horizontal rail at right angles to its direction of motion; if the height of the rail above the plane be one half the radius of the ball, shew that the latus rectum of the parabola subsequently described is one foot in length.

5. Particles are projected from the same point with equal velocities; shew that the vertices of their paths lie on an ellipse, and if they be equally elastic and impinge on a vertical wall then the vertices of their paths after impact lie on an ellipse.

6. Two equal smooth balls, whose elasticity is $\frac{1}{2}$, and which are in the same horizontal plane at a distance $2a$ apart from each other, are projected with the same horizontal velocity $\sqrt{ga}$ toward one another and are acted on by gravity only; shew that, after collision, the velocity of each ball will be $\frac{1}{2}\sqrt{5ga}$, and find the position of the vertices and foci of the subsequent paths.

7. A square is placed in a vertical plane with two sides vertical and from the foot of one of these a particle is projected with $\frac{25}{9}$ times the velocity that would be acquired in falling down it, the direction of projection being inclined at an angle $\cos^{-1}\frac{4}{5}$ to the horizontal. Find the coefficient of elasticity at the opposite vertical side in order that the particle after striking the highest side (coefficient of elasticity $\frac{9}{16}$) may return to the point of projection.

8. A perfectly elastic ball is at the focus of an elliptic billiard table; shew that the ball however struck will ultimately be moving along the major axis.

9. A ball at the focus of an ellipse, whose eccentricity is $\epsilon$, receives a blow and after one impact on the elliptic perimeter

passes through the further end of the major axis. Find the point of impact on the ellipse and shew that $\epsilon$ cannot be greater than $\dfrac{2e}{1+e^2}$.

10. The sides of a rectangular billiard table are of lengths $a$ and $b$. If a ball of elasticity $e$ be projected from a point in one of the sides, of length $b$, to strike all four sides in succession and continually retrace its path, shew that the angle of projection $\theta$ with the side is given by $ae \cot \theta = c + ec'$, where $c$ and $c'$ are the parts into which the side is divided at the point of projection.

11. Two equal balls, of radius $a$, are placed in contact on a rectangular horizontal table bounded by vertical walls so that the line joining their centres is parallel to an edge of length $h$; they are projected simultaneously towards this edge in directions inclined at angles of $45°$ with it; shew that, supposing perfect resilience, a series of collisions will occur between the balls at intervals of time equal to $\sqrt{2}\,(h-2a)/v$.

12. A smooth circular table is surrounded by a smooth rim whose interior surface is vertical. Shew that a ball of elasticity $e$ projected along the table from a point in the rim in a direction making an angle $\tan^{-1}\sqrt{\dfrac{e^3}{1+e+e^2}}$ with the radius through the point will return to the point of projection after two impacts on the rim. Prove also that when the ball returns to the point of projection its velocity is to its original velocity as $e^{\frac{3}{2}} : 1$.

13. Two equal elastic balls not in the same vertical line are dropped upon a hard horizontal plane; if the balls ever come simultaneously to a position of instantaneous rest, shew that the initial heights of the balls above the plane are as

$$(e^m - 1)^2 \text{ to } (e^n - 1)^2$$

where $m$, $n$ are positive integers and $e$ is the coefficient of elasticity.

14. A number of particles are let fall from a horizontal straight line on the convex arc of a fixed vertical parabola whose directrix is the fixed line; shew that the parabolas which they describe after impact on the curve have a common directrix at a distance $a\,(1-e^2)$ below the given line, where $4a$ is the latus rectum of the given parabola and $e$ the coefficient of restitution between the particles and the curve.

15. A number of perfectly elastic particles are projected at the same instant from a point in a horizontal plane of unlimited extent. Shew that the path of the centre of inertia of these particles will be a curve with a succession of salient points, the arcs between them being arcs of the same parabola.

16. An imperfectly elastic ball is projected with velocity $\sqrt{gh}$ at an angle $a$ with the horizon so that it strikes a vertical wall distant $c$ from the point of projection and returns to the point of projection. Shew that the coefficient of restitution between the ball and the wall is $\dfrac{c}{h \sin 2a - c}$.

17. A ball of elasticity $e$ is projected from a point in a smooth vertical wall against a parallel wall, and after $2n+1$ impacts it returns to the point of projection. If the angle of projection be $60°$ and the velocity of projection be $V$, shew that
$$V^2 = \frac{2ga}{\sqrt{3}} \frac{1-e^{2n+2}}{e^{2n+1}(1-e)},$$
where $a$ is the horizontal distance between the two walls.

18. A particle is projected from a point at the foot of one of two smooth vertical walls so that after three reflexions it may return to the point of projection and the last impact be

direct; shew that $e^3 + e^2 + e = 1$, and that the vertical heights of the three points of impact are as $e^2 : 1 - e^2 : 1$.

19. If two small perfectly elastic bodies be projected at the same instant with velocities which are as $2 \tan \beta$ to $\sqrt{1 + 4 \tan^2 \beta}$, one up a plane inclined at an angle $\beta$ to the horizon, and the other in the same vertical plane but in a direction making an angle $\theta$ with the plane such that $2 \tan \theta = \cot \beta$, they will return to the point of projection at the same instant.

20. A particle of elasticity $e$ is dropped from a vertical height $a$ upon the highest point of a plane, which is of length $b$ and is inclined at an angle $a$ to the horizon, and descends to the bottom in three jumps. Shew that

$$b = 4a \sin a\, e\, (1 + e)(1 + e + 2e^2 + e^3 + e^4).$$

21. A particle is projected from the foot of a plane inclined at an angle $\beta$ to the horizon in a direction making an angle $a$ with the inclined plane and $e$ is the coefficient of restitution between the particle and the plane. If after striking the plane $n$ times the particle rebound vertically, shew that

$$\cos(a - \beta) = \sin a \sin \beta \cdot \frac{1+e}{1-e}(1 - e^n).$$

Shew also that if $2 \tan a \tan \beta$ be $> 1 - e$, the particle will after a certain number of impacts rebound down the inclined plane.

22. A particle is projected from the foot of an inclined plane and returns to the point of projection after several rebounds, one of which is perpendicular to the inclined plane; if it take $r$ more leaps in coming down than in going up, shew that

$$\cot a \cot \theta = \{2\sqrt{1 - e^r} - 2(1 - e^r)\} \div (1 - e)\, e^r,$$

where $a$ is the inclination of the plane, $\theta$ the angle between the direction of projection and the plane, and $e$ the coefficient of restitution.

What is the condition that it may be possible to project the particle so that one of its impacts is perpendicular to the plane?

23. A particle is projected with velocity $V$ at an angle $a$ with the horizon and at the highest point of its path impinges on a wall inclined at an angle $\beta$ to the horizon; shew that after rebounding it will pass over the top of the wall if its height be

$$\frac{V^2}{2g\,\cos^2\beta}\{\sin^2 a \cos^2\beta + 4e\cos^2 a \sin^2\beta\,(\cos^2\beta - e\sin^2\beta)\},$$

and find the total range on a horizontal plane passing through the point of projection.

24. A ball, of which $e$ is the modulus of elasticity, after dropping through a height $h$ strikes at a point $A$ a plane inclined to the horizon at an angle $a$ and afterwards passes through a point $B$ in a horizontal line through $A$; find the time of moving from $A$ to $B$, and shew that the problem is impossible if $e$ be $< \tan^2 a$.

25. A ball, of elasticity $e$, is projected obliquely up an inclined plane so that the point of impact at the third time of striking the plane is in the same horizontal line with the point of projection; shew that the distances from the line of the points of first and second impact are as $1 : e$.

26. A particle, of elasticity $e$, is projected with a given velocity so that the whole distance measured up an inclined plane passing through the point of projection and described whilst parabolic motion continues is a maximum, the whole motion taking place in a vertical plane. Shew that if $a$

be the inclination of the plane and $\beta$ the angle which the direction of projection makes with the inclined plane, then $\tan 2\beta = (1-e) \cot \alpha$, and that the time that elapses before the parabolic motion ceases is to the whole time during which the particle is moving up the plane as $\cos 2\beta$ is to $\cos^2 \beta$.

27. An imperfectly elastic particle, whose coefficient of elasticity is $e$, is projected from a point in a plane inclined to the horizon at an angle $\gamma$. The direction of projection makes an angle $\beta$ with the normal to the plane and the plane through these two lines makes an angle $\alpha$ with the line of greatest slope on the inclined plane. When the particle meets the plane for the $n$th time it is in the horizontal line through the point of projection; shew that $(1-e^n) \tan \gamma = (1-e) \tan \beta \cos \alpha$, and find the distance of the particle from the point of projection.

28. A perfectly elastic particle acted on by no forces is projected from the centre of a square and rebounds from its sides; if it ever pass through an angular point, shew that the direction of projection makes with a side an angle $\tan^{-1} \dfrac{2n+1}{2m+1}$ where $m$ and $n$ are integers, and if it ever strike the middle point of a side the angle must be of the form $\tan^{-1} \dfrac{2m+1}{2n}$.

29. Three smooth billiard balls, each of radius $d$, and with coefficient of resilience unity, rest on a smooth table, their centres forming a triangle $ABC$; shew that if the ball $A$ is to cannon off $B$ on to $C$ the angle of impact at $B$ must lie between

$$B - \frac{\pi}{2} - \tan^{-1} \frac{2d \cos B}{C - 2d \sin B}$$

and $\quad B + \delta - \dfrac{\pi}{2} - \tan^{-1} \dfrac{2d \cos (B+\delta)}{C - 2d \sin (B+\delta)}$, where $\delta = \sin^{-1} \dfrac{4d}{a}$.

30. A man has to make a small billiard ball, $A$, strike another ball, $B$, after one impact on a cushion. He knows the direction in which to strike $A$ but is liable to a small angular error in striking. Shew that the positions of $B$ in which it is equally liable to be struck by $A$ lie on a curve whose polar equation referred to a certain origin can be put into the form

$$r^2 (e^2 \sin^2 \theta + \cos^2 \theta) = \text{const.},$$

where $e$ is the coefficient of restitution.

31. A perfectly elastic ball is projected horizontally from the top of a series of steps whose breadths are just greater than double their heights with a velocity equal to that which would be acquired in falling through the height of one step; shew that it falls on the next step and clears 2, 4, 6,...steps in subsequent rebounds.

32. A perfectly elastic ball is thrown into a smooth cylindrical well from a point in the circumference of the cylindrical mouth. If the ball be reflected any number of times from the surface of the cylinder, shew that the times between the successive reflections are equal. If it be projected horizontally in a direction making an angle $\dfrac{\pi}{n}$ with the tangent at the point of projection it will meet the water at the instant of the $n$th reflection provided that the velocity of projection is $nr \sin \dfrac{\pi}{n} \sqrt{\dfrac{2g}{d}}$, where $r$ is the radius and $d$ the depth of the well.

33. A hollow elliptic cylinder is placed on a horizontal table with its axis vertical. From the focus of a horizontal section a perfectly elastic ball is projected with velocity $v$ in a horizontal direction. Shew that if the particle return to the point of projection then the height of the section above the table is $2m^2 g a^2 / n^2 v^2$, where $m$, $n$ are positive integers and $2a$ the major axis of the section.

EXAMPLES ON CHAPTER VI.   235

34. Three equal balls $A$, $B$, $C$, follow one another in this order in a straight line such that $u > v > w$. The following impacts take place in order; $A$ strikes $B$, $B$ strikes $C$, and $A$ strikes $B$ again. If the coefficient of restitution be $\frac{1}{2}$ for all the balls and $4w$ be $> 25v - 21u$, shew that no more impacts will happen.

35. Two equal balls, $A$ and $B$, are lying very nearly in contact on a smooth horizontal table and a third equal ball impinges on $A$, the centres of the three balls being in the same straight line; shew that, if $e$ be $> 3 - 2\sqrt{2}$, there will be three impacts, and that the final velocity of $B$ will be to the initial velocity of the striking ball as $(1 + e)^2 : 4$.

36. Two equal balls, $A$ and $B$, of elasticity $e$, move on a horizontal plane with uniform velocity in the same straight line which is perpendicular to a given wall and $B$ impinges on the wall; shew that two impacts between $A$ and $B$ must follow and two between $B$ and the wall, and that there will be a third collision between the balls if $e$ be $< 2 - \sqrt{3}$.

37. Two particles, of masses $m$ and $m'$, are constrained to remain in a smooth circular groove, of radius $a$, which is cut in a horizontal table; $m$ is projected along the groove with velocity $u$ and impinges on $m'$; $m'$ then impinges on $m$ and so on; shew that the kinetic energy will never be so small as $\dfrac{m}{m + m'} \times$ initial kinetic energy, and that the time between the first and $n$th impacts is $\dfrac{2\pi a}{u} \dfrac{1 - e^{n-1}}{e^{n-1}(1-e)}$, where $e$ is the coefficient of restitution.

38. Two equal balls of radius $a$ are in contact and are struck simultaneously by a ball of radius $c$ moving in the direction of their common tangent; if all the balls be of the same material and the coefficient of restitution be $e$, find

the velocities of the balls after impact, and shew that the impinging ball will be reduced to rest if $2e = \dfrac{c^2 (a+c)^2}{a^3(2a+c)}$.

39. A bucket and a counterpoise connected by a string passing over a smooth pulley just balance one another and an elastic ball is dropped into the centre of the bucket from a height $h$ above it; find the time that elapses before the ball ceases to rebound, and shew that the whole descent of the bucket during this interval is $\dfrac{4mh}{2M+m} \dfrac{e}{(1-e)^2}$, where $m$, $M$ are the masses of the ball and bucket and $e$ is the coefficient of restitution.

40. If a portion of a horizontal plane begin to move upward under the action of its weight $W$ and a constant force upward, and at the same instant a particle of weight $w$ and coefficient of restitution $e$ be allowed to drop from a height $h$ above it, shew that by the time the particle is at rest relative to the plane the height travelled by the latter is less than the height through which it would have travelled had no particle dropped on it by $\dfrac{wh}{W+w} \dfrac{4e}{(1-e)^2}$.

41. A vertical shower of hail begins to fall with velocity $u$ and strikes a horizontal slab of unit area. If $M$ be the mass of the particles in a unit volume of the shower before impact, shew that at time $\dfrac{2eu}{g(1-e)}$ the pressure on the slab is $Mu^2 \dfrac{1+e}{1-e}$, where $e$ is the coefficient of elasticity.

What must be taken into account in the measure of the pressure after this time?

42. A smooth circular ring, of mass $M$ and radius $a$, rests on a smooth horizontal table, and a small spherical ball

of mass $m$ is projected from the centre of the circle with velocity $v$; shew that the whole time that elapses before the $n$th impact is $\dfrac{a}{v} \dfrac{2 - e^{n-1} - e^n}{e^{n-1} - e^n}$, where $e$ is the coefficient of restitution.

Find also the velocity of the ring and sphere after an infinite time.

43. A boy standing on a bridge lets a ball drop upon the horizontal roof of a railway carriage passing under him at the rate of 15 miles per hour. If the coefficient of elasticity between the ball and the roof be $\frac{4}{7}$ and the coefficient of friction be $\frac{1}{2}$, find the least height of the boy's hand so that the ball may rebound from the same point. If the boy's hand be at a less height than this, what will happen?

44. A particle is projected with velocity $V$ from a point in a horizontal plane in a direction making an angle $a$ with the plane. The coefficient of elasticity between the particle and the plane is $e$ and the coefficient of friction $\tan \lambda$. Shew that, if $\tan a$ be $< \dfrac{1-e}{1+e} \cot \lambda$, the particle will come to rest on the plane after a time $\dfrac{V}{g} \dfrac{\cos(a - \lambda)}{\sin \lambda}$ and that the distance described measured along the plane from the point of projection is $\dfrac{V^2}{g} \dfrac{\cos^2(a - \lambda)}{\sin 2\lambda}$.

How will the problem be modified if $\tan a$ be $> \dfrac{1-e}{1+e} \cot \lambda$?

45. Two equal spheres, $A$ and $B$, are lying in contact on a smooth table. $A$ is struck by a third equal sphere moving in a direction making an angle $a$ with $AB$; shew that the direction of motion of $A$ after the impact is inclined to $AB$ at an angle $\tan^{-1}(2 \tan a)$.

238    EXAMPLES ON CHAPTER VI.

46. A smooth ball, of mass $m'$, is lying on a horizontal table in contact with a vertical inelastic wall and is struck by another ball of mass $m$ moving in a direction normal to the wall and inclined at an angle $a$ to the common normal at the point of impact; shew that, if the coefficient of restitution between the two balls be $e$, the angle $\theta$ through which the direction of motion of the striking ball is turned is given by the equation

$$\tan(\theta + a) = -\tan a \,\frac{m' + m \sin^2 a}{em' - m \sin^2 a}.$$

47. A billiard ball in contact with a cushion is struck obliquely by another ball, the line joining their centres making an angle $\theta$ with the cushion; the coefficient of elasticity for each impact is $e$ and the moments of greatest compression are simultaneous. Shew that if the striking ball move parallel to the cushion after impact its velocity then is to that of the other as $1 - e \sec^2 \theta : 1 + e$.

48. A smooth uniform hemisphere of mass $M$ is sliding with velocity $V$ on an inelastic plane with which its base is in contact; a sphere of smaller mass $m$ is dropped vertically and strikes the hemisphere on the side towards which it is moving so that the line joining their centres makes an angle of $45°$ with the vertical; shew that, if the hemisphere be stopped dead, the sphere must have fallen through a height

$$\frac{V^2}{2g} \frac{(2M - em)^2}{(1 + e)^2 m^2},$$

where $e$ is the coefficient of restitution between the sphere and hemisphere.

49. A smooth sphere, of mass $m$, is tied to a fixed point by an inelastic string and another sphere, of mass $m'$, impinges directly on it with velocity $v$ in a direction making an acute angle $a$ with the string. Shew that $m$ begins to move with

velocity $\dfrac{m'(1+e)\sin a}{m + m'\sin^2 a}\,v$, and find the change in the velocity of $m'$.

50. Two inelastic spheres, of masses $m$ and $m'$, are in contact and $m$ receives a blow through its centre in a direction making an angle $a$ with the line of centres. Shew that the kinetic energy generated is less than if $m$ had been free in the ratio $m + m'\sin^2 a$ to $m + m'$.

51. Three equal balls of mass $m$ lie in contact on a smooth table so that their centres are at the corners of an equilateral triangle; a ball of mass $M$ impinges directly with velocity $U$ on one of the balls so that its direction of motion produced bisects the angle of the triangle; shew that the velocity of $M$ after impact is $U\dfrac{2M - 5em}{2M + 5m}$ and the velocity of the ball struck is $2U\dfrac{(1+e)M}{2M + 5m}$, assuming that the compression ends at the same moment for all the balls.

52. Two equal inelastic balls of masses $m_1$, $m_2$ are in contact, and a ball, of mass $M$, strikes both simultaneously; shew that its direction of motion is turned through an angle

$$\tan^{-1}\dfrac{(m_1 - m_2)\sin a \cos a}{M + (m_1 + m_2)\sin^2 a},$$

where $2a$ is the angle subtended at the centre of $M$ by the line joining the centres of the other two balls.

53. Three equal balls $A, B, C$ moving with the same velocity $v$ in directions inclined at angles of $120°$ to one another impinge so that their centres form an equilateral triangle. If the coefficient of elasticity between $C$ and $A$ or $B$ be $e$ and between $A, B$ be $e'$, shew that $A, B$ separate with velocity

$$\dfrac{2e' + e}{\sqrt{3}}\,v,$$

it being assumed that the compression ends at the same time for all the balls.

54.  Two smooth balls of masses $m_1$, $m_2$ are connected by an inextensible string and lie at rest at the greatest possible distance apart.  A ball of mass $M$ impinges directly on the ball $m_1$ with a velocity $V$ the direction of which makes an angle $a$ with the string.  Shew that the velocity of $M$ after the impact is

$$V \left\{ \frac{\sin^2 a}{m_1} + \frac{\cos^2 a}{m_1 + m_2} - \frac{e}{M} \right\} \div \left\{ \frac{\sin^2 a}{m_1} + \frac{\cos^2 a}{m_1 + m_2} + \frac{1}{M} \right\},$$

where $e$ is the coefficient of restitution of the balls.  Find also the velocity of the ball that is struck.

55.  A smooth fixed horizontal tube lies on a smooth table and half out of it, and just fitting it, lies a smooth ball of mass $A$.  A second ball of mass $B$ projected along the table in a direction perpendicular to the tube strikes $A$ so as to drive it into the tube.  Shew that after impact the direction of $B$'s motion makes with the common normal an angle

$$\tan^{-1} \left\{ \cot \theta \, \frac{B \cos^2 \theta + A}{B \cos^2 \theta - Ae} \right\},$$

where $\theta$ is the angle between this normal and the axis of the tube, the centres of both balls being supposed to move in the same horizontal line.

56.  An inelastic particle falls vertically upon a smooth wedge of angle $a$ which rests on a horizontal smooth table.  Shew that after impact the particle will describe a straight line with uniform acceleration and that its velocity after sliding off the wedge on to the table will be

$$\frac{\sin a \cos a}{1 + n} \sqrt{V^2 + 2ga \frac{(n+1) \operatorname{cosec} a}{1 + n \sin^2 a}},$$

where $V$ is the velocity with which the particle strikes the wedge at a distance $a$ from its foot, and $n$ is the ratio of the mass of the particle to that of the wedge.

# CHAPTER VII.

### THE HODOGRAPH AND NORMAL ACCELERATIONS.

152. In the following chapter we shall consider the motion of a particle which moves in a curve. It will be convenient, as a preliminary, to explain how the velocity, direction of motion, and acceleration of a particle moving in any manner may be marked out by means of another curve.

**Hodograph. Def.** *If a particle be moving in any path whatever and if from any point* O, *fixed in space, we draw a straight line* OQ *parallel and proportional to the velocity at any point* P *of the path, the curve traced out by the end* Q *of this straight line is called the hodograph of the path of the particle.*

153. **Theorem.** *If the hodograph of the path of a moving point* P *be drawn, then the velocity of the corresponding point* Q *in the hodograph represents, in magnitude and direction, the acceleration of the moving point* P *in its path.*

Let $P$, $P'$ be two points on the path close to one another; draw $OQ$, $OQ'$ parallel to the tangents at $P$, $P'$ and proportional to the velocities there, so that $Q$, $Q'$ are two points on the hodograph very close to one another.

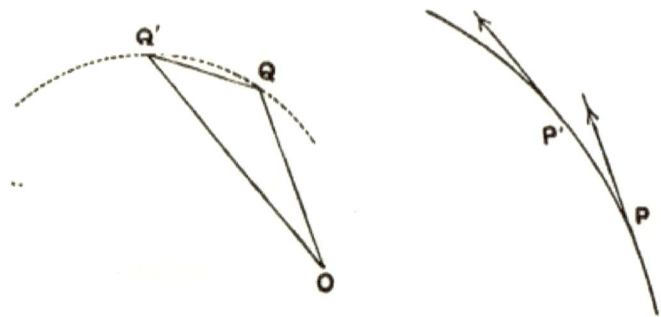

Whilst the particle has moved from $P$ to $P'$ its velocity has changed from $OQ$ to $OQ'$, and therefore, as in Art. 24, the change of velocity is represented by $QQ'$.

Now let $P'$ be taken indefinitely close to $P$ so that $QQ'$ becomes an indefinitely small portion of the arc of the hodograph.

If $\tau$ be the time of describing the arc $PP'$, then, by Art. 26, the acceleration of $P$ = limit of $\dfrac{QQ'}{\tau}$ = velocity of $Q$ in the hodograph.

Hence the velocity of $Q$ in the hodograph represents, in magnitude and direction, the acceleration of $P$ in the path.

154. *Examples.* The hodograph of a point describing a straight line with constant acceleration is a straight line which the corresponding point describes with constant velocity.

The hodograph of a point describing a circle with uniform speed is another circle which the corresponding point describes with uniform speed.

The hodograph of the path of a heavy particle projected freely is a vertical straight line which the corresponding point describes with uniform velocity.

This proposition may be easily shewn without reference to the last article. For if $v$ be the velocity at any point $P$ of a trajectory we have, by Art. 123, $v^2 = 2g \cdot SP$. But if $SY$ be the perpendicular upon the tangent at $P$ then $SY^2 = AS \cdot SP$ so that $v^2 = \dfrac{2g}{AS} SY^2$.

$\therefore v$ is proportional to $SY$.

Hence if through $S$ we draw a line $SY''$ perpendicular and proportional to $SY$ the end $Y''$ will describe the hodograph. Also since the locus of $Y$ is the tangent at the vertex of the parabola the locus of $Y''$ will be a vertical straight line.

## Normal Acceleration.

155. We have learnt from the First Law of Motion that every particle, once in motion and acted on by no forces, continues to move in a straight line with uniform velocity. Hence it will not describe a curved line unless acted upon by some external force. If it describe a curve with uniform speed there can be no force in the direction of the tangent to its path or otherwise its speed would be altered and so the only force acting on it is normal to its path. If its speed be not constant there must in addition be a tangential force.

In the following articles we shall investigate the normal acceleration of a particle moving in a given path.

**156. Theorem.** *If a particle describe a circle of radius* r *with uniform speed* v, *to shew that its acceleration is* $\dfrac{v^2}{r}$ *directed toward the centre of the circle.*

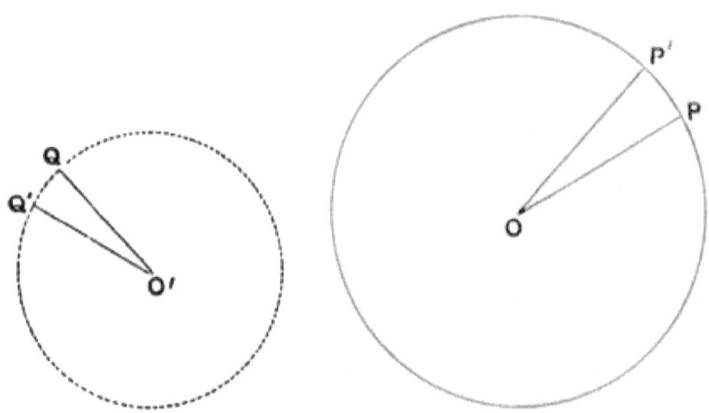

Let $P$, $P'$ be two consecutive positions of the moving point and $Q$, $Q'$ the corresponding points on the hodograph. Since the speed of $P$ is constant the point $Q$ moves on a circle whose radius is $v$; also the angle $QO'Q'$ is equal to the angle between the tangents at $P$, $P'$ and therefore to the angle $POP'$.

Hence the arc $QQ'$ : the arc $PP'$ :: $O'Q$ : $OP$ :: $v$ : $r$.

∴ the velocity of $Q$ in the hodograph : $v$ :: $v$ : $r$.

$$\therefore \text{velocity of } Q = \frac{v^2}{r}.$$

But the point $Q$ is moving in a direction perpendicular to $O'Q$ and therefore **parallel to** $PO$; also the acceleration of the point $P$ is equal to the velocity of $Q$.

Hence the acceleration of $P$ is $\dfrac{v^2}{r}$ in the direction $PO$.

## NORMAL ACCELERATION.

*Cor.* 1. If $\omega$ be the angular velocity of the particle about the centre $O$, we have $v = r\omega$, and the normal acceleration is $\omega^2 r$.

*Cor.* 2. The force required to produce the normal acceleration is $m\dfrac{v^2}{r}$ where $m$ is the mass of the particle.

*Cor.* 3. If the path of the particle be a curve whose radius of curvature at $P$ is $\rho$ and its speed be constant the same proof will apply if we suppose $O$ to be the centre of curvature at $P$; in this case the acceleration is $\dfrac{v^2}{\rho}$ along the normal at $P$.

157. In the present article we shall give a more general proof of the theorem of the preceding article in which we shall not suppose the speed in the curve to be constant, and we shall also find the tangential acceleration.

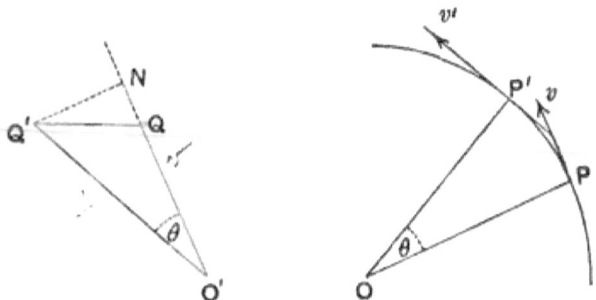

Let $P$, $P'$ be two consecutive points in the path of the particle at which the speeds are $v$ and $v'$ and let the normals at $P$, $P'$ meet in $O$. Draw $O'Q$, $O'Q'$ parallel and proportional to $v$ and $v'$ so that $QQ'$ is an element of the arc of the hodograph and therefore represents the change of velocity of the moving point in magnitude and direction.

Draw $Q'N$ perpendicular to $QN$. Then the velocity $QQ'$ is equivalent to the velocities $QN$, $NQ'$ which are parallel to the tangent and normal at $P$.

Let $\tau$ be the time of moving from $P$ to $P'$ and let the radius of curvature at $P$ be $\rho$ so that we have

$$\rho\theta = PP' = v\tau \text{ ultimately, where } \theta = POP' = QO'Q'.$$

$\therefore$ normal acceleration at $P = \mathrm{Lt}\,\dfrac{NQ'}{\tau} = \mathrm{Lt}\,\dfrac{v'\sin\theta}{\tau}$

$$= \mathrm{Lt}\,v'.\dfrac{\theta}{\tau} = \mathrm{Lt}\,\dfrac{v'.v}{\rho} = \dfrac{v^2}{\rho} \text{ ultimately,}$$

when $P$ and $P'$ are taken indefinitely close to one another.

Also the tangential acceleration at $P$

$$= \mathrm{Lt}\,\dfrac{QN}{\tau} = \mathrm{Lt}\,\dfrac{ON - OQ}{\tau} = \mathrm{Lt}\,\dfrac{v'\cos\theta - v}{\tau}$$

$$= \mathrm{Lt}\,\dfrac{v' - v}{\tau}, \text{ ultimately.}$$

Hence the tangential acceleration of the particle is equal to the rate of change of its speed.

**158.** Without the use of the hodograph a proof of the very important theorem of Art. 156 can be given as follows.

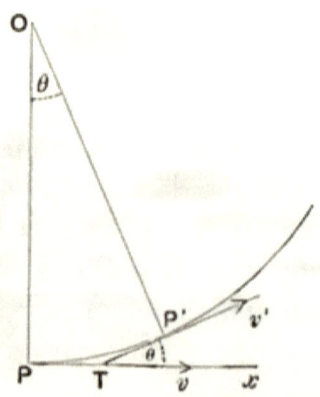

# NORMAL ACCELERATION.

Let $P'$ be a point on the path close to $P$. Draw the tangent $P'T$ at $P'$ to meet the tangent, $Px$, at $P$ in $T$, and let the normals at $P$, $P'$ meet in $O$, which, when $PP'$ becomes very small, becomes the centre of curvature at $P$.

Since the angles at $P$, $P'$ are right angles a circle will go through $O$, $P$, $T$, $P'$ and hence $P'Tx =$ supplement of $PTP' = POP' = \theta$.

Let $v$, $v'$ be respectively the velocities at $P$ and $P'$ and let $\tau$ be the time of describing the arc $PP'$.

Then in time $\tau$ a velocity parallel to $PO$ has been generated equal to $v' \sin \theta$.

Hence the acceleration in the direction $PO =$ Limit of $\dfrac{v' \sin \theta}{\tau}$

$$= \text{Limit of } v' \cdot \frac{\sin \theta}{\theta} \cdot \frac{\theta}{\text{arc } PP'} \cdot \frac{\text{arc } PP'}{\tau}.$$

But in the limit when $P'$ is taken indefinitely close to $P$ we have

$$v' = v \,; \quad \frac{\sin \theta}{\theta} = 1\,; \quad \frac{\theta}{\text{arc } PP'} = \frac{1}{\rho}\,; \quad \frac{\text{arc } PP'}{\tau} = v.$$

∴ required acceleration $= v \cdot \dfrac{1}{\rho} \cdot v = \dfrac{v^2}{\rho}$.

As before, the force in the direction of the normal to cause this acceleration must be $m \dfrac{v^2}{\rho}$.

159. The force spoken of in the preceding articles which is required to cause the normal acceleration may be produced in many ways. For example, the particle may be tethered by a string, extensible or inextensible, to a fixed point; or the force may be caused by the pressure of a material curve on which the particle moves; or, again, the force may be of the nature of an attraction such as exists between the sun and earth and which compels the earth to describe a curve about the sun.

160. When a man whirls in a circle a mass tied to one end of a string, the other end of which is in his hand, the tension of the string exerts the necessary force on the

body to give it the required normal acceleration. But, by the third law of motion, **the string exerts upon** the man's hand a **force equal and opposite to that which it exerts upon the particle**; these two **forces form the action and reaction of which Newton speaks.** It *appears* to the man that the **mass is trying to get** away from his hand. For **this reason a force equal and** opposite to the force necessary to **give the particle its normal acceleration has been called "its centrifugal force."** This is however a **very misleading term**; it seems to **imply that** the force *belongs* to the mass instead of being an external force acting on the mass. A somewhat less misleading **term is "centripetal force."** We shall avoid the use of **either term; the student who meets with them in the course of his reading will understand that either means "the force which must act on the mass to give it the acceleration normal to the curve in which it moves."**

## EXAMPLES.

1. *A particle, of mass* m, *moves on a horizontal table and is connected by a string, of length* l, *with a fixed point on the table; if the greatest weight that the string can support be that of a mass of* M *pounds, find the greatest number of revolutions per second* **that** *the particle can make without breaking the string.*

Let $n$ be the required number of revolutions so that the velocity of the mass is $n \cdot 2\pi l$.

∴ tension of the string $= m \cdot \dfrac{4\pi^2 n^2 l^2}{l}$ poundals.

Hence $Mg = 4m\pi^2 n^2 l$ so that $n = \dfrac{1}{2\pi}(Mg/ml)^{\frac{1}{2}}$.

**If the** number of revolutions were greater than this number, the **tension of** the string would be greater than the string could exert, and it would break.

EXAMPLES. 249

2. A mass held by a string of length $l$ describes a horizontal circle; if the tension of the string be three times the weight of the mass shew that the time of a revolution is $2\pi \sqrt{\dfrac{l}{3g}}$ seconds.

3. With what number of turns per minute must a mass of 10 grammes revolve horizontally at the end of a string, half a metre in length, to cause the same tension in the string as would be caused by a mass of one gramme hanging vertically?

Ans. About 13·4.

4. A body, of mass $m$, moves on a horizontal table being attached to a fixed point on the table by an extensible string whose modulus of elasticity is $\lambda$; given the original length $a$ of the string find the velocity of the particle when it is describing a circle of radius $r$.

Ans. $\sqrt{\dfrac{\lambda}{m} \dfrac{r(r-a)}{a}}$.

5. A particle whose mass is one-quarter of an ounce rests on a horizontal disc and is attached by two strings, each four feet long, to the extremities of a diameter. If the disc be made to rotate 100 times per minute about a vertical axis through its centre, find the tension of each string.

Ans. $\tfrac{25}{9}\pi^2$ poundals.

6. Two masses, $m$ and $m'$, are placed on a smooth table and connected by a light string passing through a small ring fixed to the table. If they be projected with velocities $v$ and $v'$ respectively at right angles to the portions of the string, which is initially tight, find the ratio in which the string must be divided at the ring so that both particles may describe circles about the ring as centre.

Ans. In the ratio $mv^2 : m'v'^2$.

7. A particle is moving in a circle of radius 200 feet with a speed, at a given instant, of 10 feet per second. If the speed be increasing in each second at the rate of half a foot per second find the resultant acceleration.

Ans. $\dfrac{1}{\sqrt{2}}$ . ft.-sec. unit of acceleration at an angle of 45° with the tangent to the path.

8. A skater, whose mass is **12 stone, cuts on the** outside edge a circle of **10** feet radius with uniformly decreasing speed just coming to rest **after** completing the circle in five seconds; find the direction and magnitude of his horizontal pressure on the ice when he is half way round.

*Ans.* $-\dfrac{1344}{5} \pi \sqrt{1+4\pi^2}$ poundals at an angle $\tan^{-1}(2\pi)$ with his direction of motion.

9. A wet **open umbrella is held with its handle** upright and made to rotate about it at the rate of 14 revolutions in 33 seconds. If the rim of the umbrella be a circle of one yard in diameter and its height above the ground be **four** feet, shew that the drops shaken off from the rim **meet** the ground in a circle of about five feet in diameter. If the mass of a drop be ·01 of an ounce shew that the force necessary to keep it attached to the umbrella is about ·021 of a poundal and is inclined at an angle $\tan^{-1}\frac{1}{3}$ to the vertical.

## 161. The Conical Pendulum.

If a particle be tied by a string to a fixed point $O$ and move so that it describes a circle in a horizontal plane, the string describing a cone whose axis is the vertical line through $O$, then the string and particle together are called a conical pendulum.

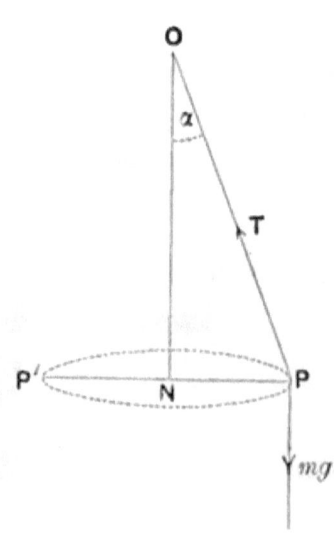

## THE CONICAL PENDULUM.

When the motion is uniform the relations between the velocity of the particle and the length and inclination of the string are easily found.

Let $P$ be the particle tied by a string $OP$, of length $l$, to a fixed point $O$. Draw $PN$ perpendicular to the vertical through $O$. Then $P$ describes a horizontal circle with $N$ as centre [dotted in the figure].

Let $T$ be the tension of the string, $\alpha$ its inclination to the vertical, and $v$ the velocity of the particle.

By Art. 156, the acceleration of $P$ in the direction $PN$ is $\dfrac{v^2}{PN}$ and hence the force in that direction is $m\dfrac{v^2}{l \sin \alpha}$.

Now the only forces acting on the particle are the tension, $T$, of the string and the weight, $mg$, of the particle.

Since the particle has no acceleration in a vertical direction we have
$$T \cos \alpha = mg \ \ldots\ldots\ldots\ldots\ldots\ldots(1).$$

Also $T \sin \alpha$ is the only force in the direction $PN$ and hence
$$T \sin \alpha = \frac{mv^2}{l \sin \alpha} \ \ldots\ldots\ldots\ldots\ldots\ldots(2).$$

From (1), (2) we have $\dfrac{v^2}{l \sin^2 \alpha} = \dfrac{g}{\cos \alpha}$.

If the particle make $n$ revolutions per second then
$$v = n \cdot 2\pi PN = 2\pi n l \sin \alpha.$$

$$\therefore 4\pi^2 n^2 l = \frac{g}{\cos \alpha}, \text{ or } \cos \alpha = \frac{g}{4\pi^2 n^2 l} \ \ldots\ldots(3).$$

Also, by (1),  $T = 4m\pi^2 n^2 l$ poundals ..............(4).

∴ tension of the string : weight of the particle

$$:: 4\pi^2 n^2 l : g.$$

The equations (3), (4) give α and $T$.

The time of revolution of the particle

$$= \frac{2\pi l \sin \alpha}{v} = 2\pi \sqrt{\frac{l \cos \alpha}{g}},$$

and therefore varies as the square root of the depth of the particle below the fixed point.

Hence it follows that for a **governor of a** steam engine **rotating 60** times per minute the height is about 9·78 inches; for one making 100 revolutions per minute the height is 3·58 inches; this latter height is too small for practical purposes except for extremely small engines.

[For many interesting details on the subject of governors of engines the student may consult Prof. Kennedy's *Mechanism*, or books on the Steam Engine.]

**162.** *Motion of bicycle rider on a circular path.* When a man is riding a bicycle on **a curved path** he always inclines his body inwards towards the centre of his path. By this means the reaction of the ground becomes inclined to the vertical. The vertical component of this reaction balances his weight, **and** the horizontal component tends towards the centre **of the** path described by the **centre** of inertia of the man and his machine and supplies the necessary normal acceleration.

**163.** *Motion of a railway carriage on a curved portion of the railway line.* When the rails are level the force to give the carriage the necessary acceleration toward the centre of curvature of its path is given by the action of the rails on the flanges of the wheels with which the rails are in contact. In order, however, to avoid the large amount of friction that would be brought into play and the consequent wearing away of the rails the outer rail is generally raised so that the floor of the train is not horizontal. The necessary inclination of the floor in order that there may be no action **on** the flanges **may be** easily found as follows.

## MOTION OF A CARRIAGE ON A CURVE.

Let $v$ be the velocity of the train, and $r$ the radius of the circle described by its centre of inertia $G$.

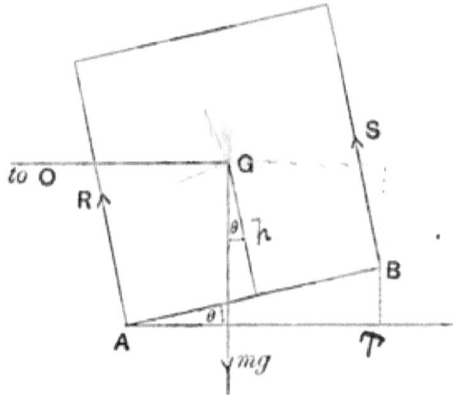

Let the figure represent a section of the carriage in the vertical plane through the line joining its centre of inertia to the centre, $O$, of the circle which it is describing and let the section meet the rails in the points $A, B$.

[The wheels are omitted for the convenience of the figure.]

Let $R, S$ be the reactions of the rails perpendicular to the floor $AB$, and let $\theta$ be the inclination of the floor to the horizon.

The resolved part, $(R+S) \sin \theta$, of the reactions in the direction $GO$ supplies the force necessary to cause the acceleration towards the centre of the curve.

$$\therefore (R+S) \sin \theta = m\frac{v^2}{r} \quad \ldots \ldots \ldots \ldots \ldots \ldots \ldots (1).$$

Also the vertical components of the reactions balance the weight,

$$\therefore (R+S) \cos \theta = mg \quad \ldots \ldots \ldots \ldots \ldots \ldots \ldots (2).$$

From (1) and (2), $\tan \theta = \dfrac{v^2}{rg}$, giving the inclination of the floor.

If the width $AB$ be given we can now easily determine the height of the outer rail above the inner.

It will be noted that the height through which the outer rail must be raised in order that there may be no pressure on the flanges depends on the velocity of the train. In practice the height is adjusted so that there is no pressure for trains moving with moderate velocities. For trains moving with higher velocities the pressure of the rails on the flanges supplies the additional force required.

## 254. ROTATING SPHERE.

**164. Rotating sphere.** *A smooth hollow sphere is rotating with uniform angular velocity $\omega$ about a vertical diameter; to shew that a heavy particle placed inside will* only remain resting against *the side of the sphere at one particular* level, *and that if the angular velocity fall short of a certain limit the particle will remain at the lowest point of the sphere.*

Let $AB$ be **the axis of rotation** of the sphere, $A$ being the highest point, and let $O$ be **the** centre; let $P$ be the position of the particle when in **relative equilibrium** and $PN$ the perpendicular on $AB$.

Now $P$ describes a circle about $N$ as centre with angular velocity $\omega$, and therefore the force towards $N$ must be $m\omega^2 \cdot PN$ or $m\omega^2 a \sin\theta$, where $a$ is the radius of the sphere and $\theta$ the angle $POB$.

The horizontal component of the **normal reaction**, $R$, at $P$ supplies this **horizontal force and the** vertical **component balances the** weight of the particle.

Hence
$$R \sin\theta = m\omega^2 a \sin\theta \quad\ldots\ldots(1),$$
$$R \cos\theta = mg \quad\ldots\ldots(2).$$

From equation (1) we have, either $\sin\theta = 0$, or $R = m\omega^2 a$.

Substituting for $R$ in (2) we have
$$\cos\theta = \frac{g}{\omega^2 a} \quad\ldots\ldots(3).$$

Hence the particle is either at the lowest point or at a point determined by equation (3).

The value of $\theta$ given by (3) is impossible unless $g < \omega^2 a$, i.e. **unless the** angular velocity **is** greater than $\left(\dfrac{g}{a}\right)^{\frac{1}{2}}$. If the angular velocity be less than this quantity **the only position of relative rest of** the particle is at the **lowest point** of the sphere.

**165. Revolving string.** *A uniform inextensible endless string revolves in the form of a circle,* **of** *radius* r, *on a horizontal smooth plane with uniform angular velocity $\omega$ about an axis through its centre perpendicular to its plane. If* m *be its mass per* **unit** *of length, to shew that its tension is* $mr^2\omega^2$.

Consider a portion, $PQ$, of the string which subtends a small angle $2\theta$ at the centre, $O$. Let the tangents at $P$, $Q$ meet in $R$ so that $QOR$, $ROP$ are each $\theta$.

REVOLVING STRING. 255

The action of the rest of the string on the element $PQ$ consists of the two tensions, each $T$, along $RP$ and $RQ$.

Their resultant is $2T \sin \theta$ in the direction $RO$.

Now the smaller the arc $PQ$ the more nearly is its motion the same as that of a particle.

The mass of $PQ$ is $m \cdot 2r\theta$.

∴ the force towards $O$ must be $2mr\theta \cdot r\omega^2$.

Hence $2T \sin \theta = 2mr^2\omega^2\theta$, or $T \dfrac{\sin \theta}{\theta} = mr^2\omega^2$.

Now let $\theta$ become indefinitely small; then in the limit $\dfrac{\sin \theta}{\theta}$ is unity, and we have
$$T = mr^2\omega^2.$$

The above investigation includes the case of a revolving fly-wheel, and gives also the *minimum* tension of a moving band which runs over a revolving shaft.

## EXAMPLES.

1. A string, of length four feet, and having one end attached to a fixed point and the other to a mass of 40 pounds revolves, as a conical pendulum, 30 times per minute; shew that the tension of the string is $160\pi^2$ poundals and that its inclination to the vertical is $\cos^{-1}\left(\dfrac{8}{\pi^2}\right)$.

2. A heavy particle which is suspended from a fixed point by a string, one yard long, is raised until the string, which is kept tight, makes an angle of 60° with the vertical and is then projected horizontally in the direction perpendicular to the vertical plane through the string; find the velocity of projection so that the particle may move in a horizontal plane.

*Ans.* 12 feet per second.

3. A particle, of mass $m$, is fastened by a string, of length $l$, to a point at a distance $b$ above a smooth table; if the particle be made to revolve on the table $n$ times per second find the pressure on the table. What is the greatest value of $n$ so that the particle may remain in contact with the table?

*Ans.* $m(g - 4\pi^2 n^2 b)$ poundals; $\dfrac{1}{2\pi}\sqrt{\dfrac{g}{b}}$.

256   EXAMPLES.

4. A particle is attached to a point $A$ by an elastic string, whose **modulus** of elasticity is twice the weight of the particle and whose natural **length is** $l$, and whirled so that the string describes the surface of a cone **whose** axis is the vertical line through $A$. If the distance below $A$ **of the** circular path during steady motion **be** $l$, shew that the velocity of **the** particle must be $\sqrt{3gl}$.

5. Two **masses,** $m$ and $m'$, are connected by **a** string, of length $c$, which passes **through a** small ring; **find** how many revolutions per second the smaller mass, $m'$, must make, as a conical pendulum, in order that the **greater** mass may rest at **a** distance $a$ from the ring.

*Ans.* $\dfrac{1}{2\pi}\sqrt{\dfrac{m}{m'}\dfrac{g}{c-a}}$.

6. A railway carriage, of mass 2 tons, is moving at the rate of 60 miles per hour on a curve of 770 feet radius; if the outer rail be not raised above the inner, shew that the lateral pressure on the rails is equal **to the** weight of about 1408 pounds.

7. **A** train **is** travelling at the rate of 40 miles per hour on a **curve the radius** of which is a quarter **of a** mile. **If the distance between the** rails be five feet find how much the outer rail must be raised above the inner so that there may be no lateral pressure on the rails.

*Ans.* About 4·9 inches.

8. A mass is hung from the roof of a railway carriage by means of a string, six feet long; **shew that, when the train is** moving **on** a curve of radius 100 yards at the rate of 30 miles per hour, the mass will move from the vertical through a distance of 1 foot 2¼ inches approximately.

9. A smooth right cone, whose vertical angle is $2a$, with axis vertical and vertex downward is made to revolve about its **axis** with uniform angular **velocity** $\omega$; **find** where **a** particle **must be placed** on the inner surface **of the cone** so that it may be in relative **equilibrium.**

*Ans.* Its distance from **the** axis of the cone must be $g \cot a/\omega^2$.

10. A smooth parabolic tube, whose latus rectum is $4a$ and vertex downward, revolves uniformly about its axis, which is vertical. Shew **that, if** the angular velocity be $\sqrt{\dfrac{g}{2a}}$, a particle will rest anywhere, whilst if the angular velocity differ from this quantity it will rest in no position except the lowest point of the tube.

11. A uniform circular string, of radius 10 feet and mass 1 lb., rotates uniformly about its centre 10 times per second; shew that the string will break unless its breaking tension is greater than the weight of 196 lbs.

12. A thin circular hoop, six feet in diameter, is revolving in its own plane about its centre. The ultimate tensible strength of iron is 60000 lbs. per square inch of section and its density is 480 lbs. per cubic foot. Find the greatest number of revolutions per second that the hoop can make without falling to pieces.

*Ans.* $40\sqrt{10}/\pi$.

166. The general case of the motion of a particle constrained to move on a given curve under any given forces is beyond the scope of the present book; so also is the motion of a particle constrained to move under gravity on a given curve.

There is one proposition, however, relating to the motion of a particle under gravity which we can prove in an elementary manner and which is extremely useful for determining many of the circumstances of the motion.

167. **Theorem.** *If a particle slide down an arc of any smooth curve in a vertical plane and if* u *be the initial velocity and* v *its velocity after sliding through a vertical distance* h, *to shew that* $v^2 = u^2 + 2gh$.

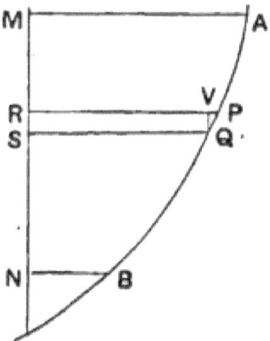

Let $A$ be the point of the curve from which the particle starts and $B$ the point whose distance from $A$, measured vertically, is $h$. Draw $AM$, $BN$ horizontal to meet any vertical line in $M$ and $N$.

Let $P$, $Q$ be two points on the curve very close to one another and draw $PR$, $QS$ perpendicular to $MN$. Then $PQ$ is very approximately a small portion of a straight line. Draw $QV$ vertical to meet $PR$ in $V$.

The acceleration at $P$ along $PQ$ is $g \cos VQP$ and hence, if $v_P$, $v_Q$ be the velocities at $P$ and $Q$, we have

$$v_Q^2 = v_P^2 + 2g \cos VQP \cdot PQ = v_P^2 + 2g \cdot VQ.$$

$$\therefore v_Q^2 - v_P^2 = 2g \cdot VQ, \quad \text{i.e. the change}$$

in the square of the velocity is due to the vertical height between $P$ and $Q$. Since this is true for every element of arc it is true for the whole arc $AB$. Hence the change in the square of the velocity in passing from $A$ to $B$ is that due to the vertical height $h$ so that $v^2 = u^2 + 2gh$.

**168.** The theorem in the preceding article may be deduced directly from the Principle of the Conservation of Energy.

For, since the curve is smooth, the reaction of the arc is always perpendicular to the direction of motion of the particle. Hence, by Art. 89, no work is done on the body by the pressure of the curve. The only force that does work is the weight of the particle.

Hence, since the change of energy is equal to the work done, we have

$$\tfrac{1}{2}mv^2 - \tfrac{1}{2}mu^2 = \text{work done by the weight} = mgh.$$

$$\therefore v^2 = u^2 + 2gh.$$

169. If instead of sliding *down* the smooth curve the particle be started along it with velocity $u$ so as to move *upwards*, the velocity $v$ when its vertical distance from the starting point is $h$ is, similarly, given by the equation

$$v^2 = u^2 - 2gh.$$

Hence the velocity of the particle will not vanish until it arrives at a point of the curve whose vertical height above the point of projection is $\dfrac{u^2}{2g}$.

It will be noticed that the height to which the particle will ascend is independent of the *shape* of the constraining curve, nor need it continually ascend. The particle may first ascend, then descend, then ascend again, and so on; the point at which it comes to rest finally will be at a height $\dfrac{u^2}{2g}$ above the point at which its velocity is $u$.

It follows that if a particle slide from rest upon a smooth arc it will come to rest when it is at the same vertical height as the starting point. An approximate example is the Switch-back railway in which the car almost rises to the same height as that of the point at which it started. The slight difference between theory and experiment is caused by the resistance of the air and the friction of the rails which, although small, are not quite negligible.

The expression for the velocity when the particle is at a vertical distance $h$ from the starting point is the same whether the particle be at that instant ascending or descending.

170. The theorem is true not only of gravity but in any case of the motion of a particle on a smooth curve under the action of a constant force in a constant direc-

tion, e.g. in the case of motion on a smooth inclined plane, if we substitute for "$g$" the acceleration caused by the forces. It is also true if we substitute for the constraining curve an inextensible string fastened to a fixed point, or a weightless rod which is always normal to the path of the particle.

We cannot, in general, find the *time* of describing any given arc without the use of the Differential Calculus.

**171. Newton's Experimental Law.** By using the theorem of Art. 167 we can shew how Newton arrived at his law of impact as enunciated in Art. 141.

We suspend two spheres, of small dimensions, by parallel strings whose lengths are so adjusted that when hanging freely the spheres are just in contact with their centres in a horizontal line.

One ball, $A$, is then drawn back, the string being kept tight, until its centre is at a height $h$ above its original position and then allowed to fall. Its velocity $v$ on hitting the second ball $B$ is $\sqrt{2gh}$.

Let $v'$, $v''$ be the velocities of the spheres immediately after the impact, and $h'$, $h''$ the heights to which they rise before again coming to rest so that

$$v' = \sqrt{2gh'}, \text{ and } v'' = \sqrt{2gh''}.$$

The sphere $A$ may either rebound, remain at rest, or follow after $B$.

Taking the former case, the relative velocity after impact is $-(v' + v'')$ or $-\sqrt{2g}(\sqrt{h'} + \sqrt{h''})$.

Also the relative velocity before impact was $\sqrt{2g} \cdot \sqrt{h}$.

We should find that the ratio of $(\sqrt{h'} + \sqrt{h''})$ to $\sqrt{h}$ would be the same whatever be the value of $h$ and the ratio of the mass of $A$ to that of $B$, and that it would depend simply on the substances of which the masses consist.

We have only considered one of the simpler cases. By carefully arranging the starting points and the instants of starting from rest both spheres might be drawn aside and allowed to impinge so that at the instant of impact both were at the lowest points of their path. The law enunciated by Newton would be found to be true in all cases.

172. **Motion on the outside of a vertical circle.** *A particle slides from rest at the highest point down the outside of the arc of a smooth vertical circle; to shew that it will leave the curve when it has described vertically a distance equal to one third of the radius.*

Let $O$ be the centre and $A$ the highest point of the circle. Let $v$ be the velocity of the particle when at a point $P$ of the curve, $R$ the pressure of the curve there, and $r$ the radius of the circle. Draw $PN$ perpendicular to the vertical radius $OA$ and let $AN = h$.

Then $\quad\quad\quad\quad v^2 = 2g \cdot AN = 2gh.$

The force along $PO$ is $\quad mg \cos \theta - R.$

But the force along $PO$ must, by Art. 156, be $m \cdot \dfrac{v^2}{r}$.

$$\therefore m \frac{v^2}{r} = mg \cos \theta - R,$$

$$\therefore R = m \left[ g \cos \theta - \frac{v^2}{r} \right] = m \left[ g \frac{r-h}{r} - \frac{2gh}{r} \right]$$

$$= mg \frac{r - 3h}{r}.$$

Now $R$ vanishes and changes its sign when $3h = r$, or when $h = \dfrac{r}{3}$. The particle will then leave the curve and describe a parabola freely; for to make it continue on the circle the pressure $R$ would have to become a tension; but this is impossible since the curve cannot *pull* the particle.

*Cor.* If the curve, instead of being a circle, be any other curve in a vertical plane the particle will leave it similarly at a depth $h$ where

$$2h = \rho \cos \theta,$$

$\rho$ being the radius of curvature at that depth and $\theta$ its inclination to the vertical.

# MOTION IN A VERTICAL CIRCLE.

**173. Motion in a vertical circle.** *A particle, of mass* m, *is suspended by a string, of length* r, *from a fixed point and hangs vertically. It is then projected with velocity* u *so that it describes a vertical circle; to find the tension and velocity at any point of the subsequent motion and to find* **also** *the conditions that it may* (α) *make complete revolutions,* (β) *oscillate, and* (γ) *cease moving in a* **circle**.

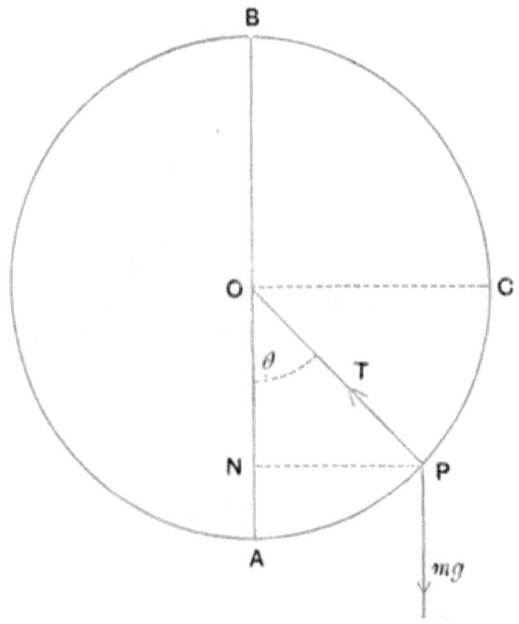

**Let O be the** point to which the string is attached, $OA$ the vertical through $A$ and $OC$ the horizontal radius.

**Let $v$ be the velocity** at any point $P$ **of** its path and $T$ the tension of the string there.

Let $PN$ be drawn perpendicular to $OA$, let $AN = h$, and let the angle $POA$ be $\theta$.

Then, by Art. 167, $\qquad v^2 = u^2 - 2gh \dots\dots\dots\dots\dots\dots\dots\dots\dots(1)$.

Also, by Art. 156,

$m \dfrac{v^2}{r} =$ force at $P$ along the **normal** $PO$ to the path of the particle.

$$\therefore m \frac{v^2}{r} = T - mg \cos\theta = T - mg \frac{r-h}{r},$$

$$\therefore T = m \frac{v^2 + g(r-h)}{r} = m \frac{u^2 + g(r-3h)}{r} \dots\dots\dots\dots(2).$$

MOTION IN A VERTICAL CIRCLE.

These two equations give **the velocity** and tension **at any point.**

(α) The particle will make complete revolutions if **the velocity do not** vanish before **it reaches** the highest point $B$ of **the path and if,** in addition, the value of the tension given by (2) do not become negative.

This latter **condition is** necessary; for otherwise, in order that the circular **motion might** continue, the *pull* of the string would **have to** change into a *push* **and this** is impossible in the case of a string.

It **follows from** (2) that $T$ diminishes **as** $h$ increases; hence if it be positive at the highest point $B$, where $h = 2r$, it is positive everywhere.

This will be the **case if** $u^2 + g(r - 6r)$ be positive, or if $u^2$ be $> 5gr$.

Also, if $u^2$ be $> 5gr$, the value of $v$ given by (1) does not vanish anywhere **and the particle makes complete revolutions.**

If $u^2$ be $< 5gr$ the tension vanishes before the particle reaches the highest point and two cases arise.

The particle may **rise to a certain** height and then retrace its path, or it may leave the circular path altogether.

The first of these two cases arises when the velocity $v$ vanishes whilst the tension still **remains positive; the latter case when the** tension vanishes before **the velocity.**

(β) From **(1) the velocity** vanishes at a height $u^2/2g$ above the lowest point, and from **(2) the** tension vanishes at a height $(u^2 + gr)/3g$.

Hence the velocity vanishes before the tension if

$u^2/2g$ be $< (u^2 + gr)/3g$, i.e. if $u^2$ be $< 2gr$,

i.e. if the velocity **of projection be insufficient to take the particle as high as the centre** $O$.

**Hence the particle will oscillate if** $u^2$ be $< 2gr$.

If $u^2 = 2gr$ the tension will **vanish at the same time as the velocity** but, as it has not become negative, **the motion** will still be oscillatory.

(γ) If $u^2$ be $> 2gr$, then $u^2/2g$ is $> (u^2 + gr)/3g$, and the tension will vanish before the velocity at a height $(u^2 + gr)/3g$, which is greater than $r$. Hence the circular motion **will cease at some point of** the path between $C$ and $B$; the string will become slack **and the particle will describe a parabola freely.**

*To sum up the results of this investigation.*

If $u$ be $< \sqrt{2gr}$ the particle oscillates in an **arc less than a** semicircle.

If $u = \sqrt{2gr}$ the arc of oscillation is a semi-circle.

If $u$ be $> \sqrt{2gr}$ and $< \sqrt{5gr}$ the particle ceases describing the circle somewhere between $C$ and $B$.

If $u=\sqrt{5gr}$ the particle *just* makes complete revolutions.

The results of this article apply without any alteration if the particle move on the *inside* of a **metal curve in the shape of the above** circle. We have only to substitute **the pressure of the curve for the tension of the string.**

*Cor.* The maximum value of the tension is at the lowest point and its value there is $m(u^2+gr)/r$; also if the particle makes complete oscillations $u \not< \sqrt{5gh}$.

∴ the tension is not $< 6mg$, and hence the breaking tension of the string must be at least six times the weight of the particle.

**174.** If the motion be inside a metal tube in the shape of a circle the investigation of the motion is **similar** to that of the last article; in this case, however, the pressure can change its sign; for the particle may pass from contact with the one **surface of** the tube to contact with the other. Hence the particle will make complete revolutions **if the velocity of projection** be sufficient to carry it to the highest point, i.e. if $u$ be $\not< \sqrt{4gr}$; **otherwise it** will oscillate.

Ex. 1. A particle slides down **the arc of a vertical** circle; shew that its velocity at the lowest point **varies as the chord** of the arc of descent.

Ex. 2. A heavy particle is attached by a string, 10 feet long, to a fixed point and swung round in a vertical circle. Find the tension and velocity at the lowest point of the circle so that the particle may just make complete revolutions.

*Ans.* Six times the weight of the particle; 40 feet per second.

Ex. 3. If a particle move in a smooth **vertical** circular tube, of radius $a$, with velocity **due to a** height $h$ above the lowest point of the circle, prove **that the pressure at** any point will be proportional to the depth below a horizontal line at a height $\dfrac{2h+a}{3}$ above the lowest point.

Ex. 4. A particle is **projected along the inside** of a smooth vertical circle, of radius $a$, from **the lowest point**; find the velocity of projection so that after leaving the circle the particle may pass through the centre.

*Ans.* $(\sqrt{3}+1)\sqrt{\tfrac{1}{2}ga}$.

APPARENT VALUE OF GRAVITY. 265

Ex. 5. A heavy particle is fastened to the middle point, $C$, of a string $ACB$, two yards in length, the ends of which, $A$ and $B$, are fastened to fixed points in the same horizontal line at a distance $3\sqrt{3}$ feet from each other. The particle, when hanging at rest, has given to it a velocity of 16 feet per second perpendicular to the plane $ACB$; shew that it will completely describe a vertical circle and that the greatest and least tensions of the string will be respectively nineteen-thirds and one-third of the weight of the particle.

**175.** *Effect of the rotation of the earth on the apparent weight of a body.* The earth rotates once per day on its axis so that any particle $P$ on its surface describes a circle round the foot $N$ of the perpendicular drawn from $P$ to the axis. Hence there must be some force in the direction $PN$. Now the forces acting on a particle resting on the earth's surface are (1) the attraction of the earth on the particle, and (2) the pressure of the ground on it. The resultant of these two forces must be in the direction of the line $PN$. This pressure of the earth on the body is the apparent value of gravity and is the same as the tension of the string which would support the body if hanging freely.

**176.** *Assuming the earth to be a sphere of radius a feet, to find the apparent value and direction of the attraction of the earth at any point of its surface taking into consideration the revolution of the earth about its axis.*

Let $P$ be a point in latitude $\lambda$, $PN$ the perpendicular on the axis, $OA$, of the earth, and $APB$ the meridian through $P$, so that the angle $POB$ is $\lambda$. Also let the angular velocity of the earth be $\omega$.

Let $m$ be the mass of a particle at $P$, $mg$ the attraction of the earth on it directed towards the centre $O$, and let $Y$, $X$ be the components of the pressure of the earth on the particle along and perpendicular to the radius $OP$ respectively.

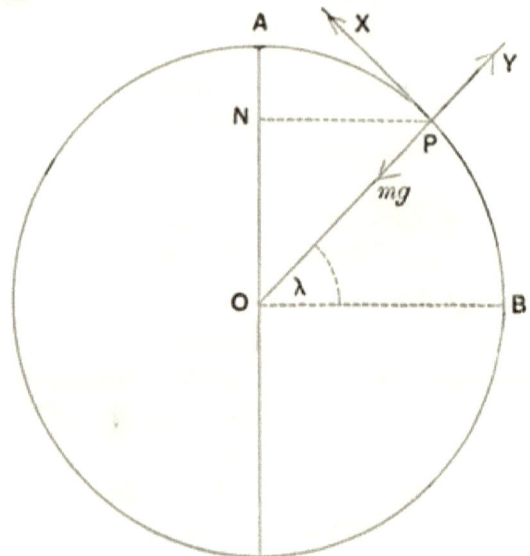

Since $P$ describes, with uniform angular velocity $\omega$, a circle whose centre is $N$, its only acceleration is $\omega^2 . PN$ directed along $PN$.

Hence resolving the forces $X$, $Y$, and $mg$ along and perpendicular to $PN$ we have

$X \sin \lambda + (mg - Y) \cos \lambda = m\omega^2 . PN = m\omega^2 a \cos \lambda \ldots (1),$

$X \cos \lambda - (mg - Y) \sin \lambda = 0 \ldots\ldots\ldots\ldots\ldots\ldots\ldots (2).$

Solving these equations we have

$X = m\omega^2 a \cos \lambda \sin \lambda,$ and $Y = m(g - \omega^2 a \cos^2 \lambda).$

Assuming as a rough approximation that the earth turns on its axis once in 24 hours and that the radius of the earth is 4000 miles, we have

$$\frac{\omega^2 a}{g} = \left(\frac{2\pi}{24 \times 60 \times 60}\right)^2 \times \frac{4000 \times 1760 \times 3}{32 \cdot 2} = \text{about } \frac{1}{289}.$$

Hence $\dfrac{\omega^2 a}{g}$ is a small quantity and so its square may be neglected.

Let the resultant of $X$ and $Y$ be $G$ at an angle $\theta$ with the radius $OP$.

Then
$$G = \sqrt{X^2 + Y^2}$$
$$= mg\left\{\left(1 - \frac{\omega^2 a}{g}\cos^2\lambda\right)^2 + \left(\frac{\omega^2 a}{g}\cos\lambda\sin\lambda\right)^2\right\}^{\frac{1}{2}}$$
$$= mg\left(1 - \frac{\omega^2 a}{g}\cos^2\lambda\right),$$

(by the Binomial Theorem, neglecting the square of $\dfrac{\omega^2 a}{g}$),

$$= mg - m\omega^2 a \cos^2\lambda.$$

Also $\tan\theta = \dfrac{X}{Y} = \dfrac{\omega^2 a \cos\lambda \sin\lambda}{g - \omega^2 a \cos^2\lambda} = \dfrac{\omega^2 a}{g}\sin\lambda\cos\lambda.$

Hence the effect of the rotation of the earth is to diminish the apparent weight by $m\omega^2 a \cos^2\lambda$, and to turn its direction through an angle

$$\tan^{-1}\left(\frac{\omega^2 a}{g}\sin\lambda\cos\lambda\right).$$

At the equator the apparent weight is less than the real weight by $m\omega^2 a$; if we take accurate values of the quantities involved this diminution is found to be about $\dfrac{1}{293}$ of the real weight.

*Cor.* The apparent pressure at the earth's surface at the equator would just vanish (i.e. the attraction of the earth would be just sufficient to keep a particle rotating with the earth) if $Y$ were zero there, i.e. if the angular velocity $\omega$ of the earth were $\sqrt{\dfrac{g}{a}}$. If the angular velocity of the earth were to exceed this quantity particles would not remain in contact with the earth unless compelled to do so.

This limiting angular velocity is roughly

$$\sqrt{\frac{32}{4000 \times 1760 \times 3}} \text{ or } \frac{1}{100\sqrt{66}}.$$

The time of the earth's revolution would then be

$$2\pi \div \frac{1}{100\sqrt{66}}$$

seconds or about 1 hour 25 minutes.

Hence if the earth were to rotate about **17** times faster than at present bodies at the equator would leave its surface.

**Ex.** If the earth were set to revolve on its axis in half a day, nothing else being altered, find what would be the apparent weight of a mass of 100 pounds, as tested by a spring balance, supposing the balance to be graduated to shew the weight of pounds at the equator, and assuming that the earth's radius is $21 \times 10^6$ feet and that the value of $g$ is 32·09 at the equator with the actual angular velocity.

*Ans.* Wt. of 98·96 lbs.


## EXAMPLES. CHAPTER VII.

1. A railway carriage is travelling on a curve of radius $r$ with velocity $v$; if $h$ be the height of the centre of inertia of the carriage above the rails and $2a$ the distance between them, shew that the weight of the train is divided between the inner and outer rails in the ratio of $gra - v^2h$ to $gra + v^2h$, and hence that the carriage will upset if $v$ be $> \sqrt{\dfrac{gra}{h}}$.

# EXAMPLES ON CHAPTER VII.

2. A particle, of mass $m$, is at the centre of a fine elastic string, of natural length $2a$, the ends of which are held in a man's hands at a distance $2a$ apart. If the mass be then swung round with angular velocity $\omega$, shew that the angle which the portions of the string make with the line joining the hands is $2\sin^{-1}\left(\dfrac{\omega}{2}\sqrt{\dfrac{ma}{\lambda}}\right)$, where $\lambda$ is the modulus of elasticity of the string.

3. A bead is strung on a thread whose two extremities are fixed; shew that if it be projected so as to move on a horizontal plane passing through the two points then the tension of the string varies inversely as the product of the two portions of the string joining it to the fixed points.

4. A leather belt, whose mass is 3 lbs. per foot of its length, is running over a pulley with a velocity of 100 feet per second. Find the least tension permissible in the belt and shew that it is independent of the radius of the pulley.

5. A circular elastic band is placed about a wheel whose circumference is twice the natural length of the string and the wheel revolves with constant angular velocity; find the pressure of the band on the wheel.

6. Shew that at the equator a shot fired westward with a velocity of 8333 metres per second, or eastward with 7407 metres per second, would, if unresisted, move horizontally round the earth in one hour and twenty minutes and one hour and a half respectively.

7. A man stands at the North Pole and whirls a mass of 24 lbs. Troy on a smooth horizontal plane by a string, one yard long, at the rate of 100 turns per minute; he finds that the difference of the forces which he has to exert according as he whirls one way or the other is roughly equal to the weight of 39 grains; find the period of rotation of the earth.

8. A railway train, of mass $m$, is travelling on a smooth line with velocity $v$ along a parallel of latitude in latitude $\lambda$; shew that the difference in the normal pressure according as it is going east or west is $4mv\omega \cos \lambda$ poundals, where $\omega$ is the angular velocity of the earth.

9. If the earth were to rotate so fast that particles on its surface at the equator were just on the point of leaving it, prove that the greatest inclination of the plumb-line at any point to the **radius of** the earth through the **point** would be $\tan^{-1} \frac{1}{2}$.

10. A string, of length $l$, has its ends fastened to two points, $A$ and $B$, in the same vertical line and a bead $P$ on the string rotates about $AB$ with a uniform angular velocity $\omega$. If $\omega$ be less than $\{2gl/(l^2-a^2)\}^{\frac{1}{2}}$ shew that it will hang vertically, and that $BP$ will be horizontal if $\omega$ be equal to $l\{2g/a(l^2-a^2)\}^{\frac{1}{2}}$.

11. A string, of length $a$, is fastened to a point $A$ and carries a mass $P$. If $P$ be projected vertically upwards when $AP$ is below the horizontal line and makes an angle $a$ with it, find the least velocity of projection that $P$ may ultimately describe a circle.

12. A particle hangs vertically by a string and is projected horizontally and rises to $P$; it then leaves circular motion and commences circular motion again at $Q$; prove that $PQ$ and the tangent at $P$ make equal angles with the vertical.

13. A string passing through a **small hole in** a smooth horizontal table has a small sphere, of mass $m$, attached to each end of it; the upper sphere revolves in a circle on the table when suddenly it strikes an obstacle and loses half its velocity; find what diminution must be made in the mass of the lower sphere so that the upper one may continue rotating in a circle.

EXAMPLES ON CHAPTER VII. 271

14. A string $PAQ$ passes through a hole $A$ in a smooth table the portion $AP$ lying on the table and $AQ$ being at an angle of $45°$ to the vertical and below the table, so that $P$ and $Q$ are in the same vertical line. If masses be attached at $P$ and $Q$ and, the strings being stretched, be each projected horizontally, find the velocity of projection and the ratio of the masses so that the plane $PAQ$ may always be vertical and the angle $PAQ$ always $45°$. If the string be four feet in length find the time of revolution.

15. Two particles, of masses $m$ and $m'$, are connected by an inextensible string of length $a$ and lie on a smooth table at a distance $a$ from each other. The particle of mass $m$ is projected at right angles to the string with velocity $v$. Shew that the tension of the string is always $\dfrac{mm'}{m+m'}, \dfrac{v^2}{a}$.

16. A particle is projected horizontally from the highest point of a parabola whose axis is vertical and vertex upwards; if the particle ever leave the curve at all it will do so at the instant of projection.

17. A smooth parabolic curve is fixed with its plane vertical and axis horizontal; shew that a heavy particle placed on the arc at a height above the axis equal to the latus rectum will run off the curve on arriving at the end of the latus rectum and then describe a parabola having the same latus rectum as the given arc.

18. A bead slides on a wire bent into the form of a parabola whose axis is vertical and vertex upwards; if the bead be just displaced from its position of equilibrium, shew that in any subsequent position the pressure on the curve varies as the curvature of the curve, and the velocity of the bead varies as its distance from the axis of the parabola.

19. Particles start from all points of the arc of a parabola, whose latus rectum is $4a$ and whose axis is vertical, and slide down to the vertex arriving there simultaneously. They then proceed to fall freely; shew that after a time $\sqrt{\dfrac{2a}{g}}$ each is at the same distance from the axis that it was initially.

20. An ellipse is placed with its minor axis vertical and a smooth particle slides down it starting from rest at the highest point and leaves the curve at the point whose eccentric angle is $\phi$. **Shew that the eccentricity of** the ellipse is

$$\frac{1}{1-\sin\phi}\sqrt{\frac{2-3\sin\phi}{2+\sin\phi}}.$$

21. An elliptic wire of eccentricity $\sqrt{\dfrac{2}{3}}$ is placed in a vertical plane with its major axis inclined to the horizon at an angle of $60°$. If a small ring be allowed to slide along the wire from the higher extremity of the major axis shew that its velocities at the two ends of the minor axis are as $\sqrt{2} : 1$.

22. An ellipse in a vertical plane has its major axis horizontal and a particle is constrained to describe the circle of curvature at its lowest point starting from rest at the highest point of the circle; at the lowest point it is gently guided inside the ellipse; shew that the particle will completely describe the ellipse if its eccentricity be greater than $\frac{1}{2}$.

23. Two equal particles at rest are connected by a fine inextensible string of length $3a$. One particle is free to move on a horizontal table in a smooth parabolic groove whose latus rectum is $4a$. The string passes through a hole in the table at the centre of curvature of the vertex of the groove and initially the second particle is just at the level of the table.

It then falls freely under gravity; shew that the first particle, when it is passing through the vertex of the groove, exerts no pressure on the sides of the groove.

24. Shew that a particle of mass $m$ may describe a parabola, whose focus is $S$ and vertex $A$, with uniform velocity $V$ under the action of two equal forces, one attractive towards $S$ and the other repulsive and perpendicular to the directrix, provided that the forces are each inversely proportional to the distance of the particle from the focus and are each equal to $m\dfrac{V^2}{4SA}$ when the particle is at the vertex.

25. A particle is describing uniformly a horizontal circle within a hollow bowl in the shape of a prolate spheroid whose axis is vertical; shew that the time of a complete revolution varies as the square root of the depth of the particle below the centre of the spheroid.

26. A particle slides from rest at the vertex down the arc of a smooth vertical circle; find the equation of its hodograph.

27. A ball of mass $m$ is just disturbed from the top of a smooth circular tube in a vertical plane and falls impinging on a ball of mass $2m$ at the bottom, the coefficient of restitution being $\frac{1}{2}$; find the height to which each ball will rise in the tube after a second impact.

28. A particle at the end of a fine string just makes complete revolutions in a vertical circle. If the particle were five times as heavy as it is the string would just bear the greatest strain. At the highest point of its path another particle of five times its mass and moving in the same circle with five times its velocity overtakes it; given that the string is now on the point of breaking when the tension is greatest shew that the modulus of elasticity of the particles is equal to $\frac{1}{5}$.

L. D. 18

EXAMPLES ON CHAPTER VII.

29. A light inextensible string has one extremity fixed, a **heavy** particle attached at the other extremity, and another **heavy** particle attached at an intermediate point; the particles describe horizontal circles with the same angular velocity $\omega$. Shew that the inclinations of the two parts of the string to the vertical are $\cot^{-1} g/\omega^2 r_1$ and $\cot^{-1} g/\omega^2 r_2$, where $r_1$ and $r_2$ are the radii of the circles described by the centre of inertia of the two particles and the lower particle respectively.

30. Watt's Governor consists of a rhombus of jointed rods attached **at** one angle to the top of a vertical revolving shaft and at the opposite angle to a collar free to slide up and down the shaft and carrying two heavy masses at the angles, each equal to $M$. Supposing the masses of the rods to be neglected in comparison with $M$, find the angle to which the rods open as the shaft revolves and find the change produced in this angle by attaching **a load** $M'$ **to** the collar.

31. Two particles, of masses $m$ and $m'$, are fixed to the ends of a weightless rod, of length $2l$, which is freely moveable about its middle point. Shew that the inclination $a$ of the rod to the vertical when the particles are moving with uniform angular velocity $\omega$ about a vertical axis through the centre of the rod is given by the equation $(m + m') \omega^2 l \cos a = (m - m') g$.

32. Assuming that the planets **move** in circles about the sun **as** centre under **an** attraction **to** the sun which varies inversely as the square **of** the distance, shew that the squares of their times of revolution vary as the cubes of their distances.

33. A gun is suspended freely by two equal parallel cords so that its inclination to the horizon is $a$ and a shot whose **mass is** $\dfrac{1}{n}$-th that **of the** gun is fired from it. If $h$ be the

height through which the gun recoils shew that the range of the shot on the horizontal plane through the muzzle is

$$4n(1+n)h\tan a.$$

34. A loaded cannon is suspended from a fixed horizontal axis and rests with its axis horizontal and perpendicular to the fixed axis, the supporting ropes being equally inclined to the vertical; if $v$ be the initial velocity of the ball whose mass is $\frac{1}{n}$th that of the cannon and $h$ be the distance between the axis of the cannon and the fixed axis of support, shew that, when the cannon is fired off, the tension of each rope is immediately altered in the ratio $v^2 + n^2gh : n(n+1)gh$.

If the cannon were supported in a gun-boat in the manner described with its axis in the direction of the boat's length what would be the effect of firing it?

35. Two equal particles are connected by a string, of length $\pi a$, and rest symmetrically over a smooth vertical circle of radius $a$. If they be just displaced from the position of equilibrium, shew that the upper particle will leave the arc when the radius to it from the centre makes an angle $\theta$ with the vertical given by $\left(\theta + \dfrac{\pi}{2}\right) = 2\cos\theta$.

36. A lamina in the form of a regular hexagon, whose side is $a$, is placed on a smooth table and fixed to the table. A string, whose length is equal to the perimeter of the lamina, is wound round it, one end being attached to an angular point, and carries a mass $m$ attached to its other end. If the particle be projected, with velocity $u$, horizontally at right angles to the string, find the time that elapses before the string is again wound up and determine the greatest and least tensions of the string.

37. A prism (whose transverse section is a regular hexagon) is held with its axis horizontal and a string equal in length to the perimeter of the hexagon with a heavy particle at one end is suspended by the other end from a point in the lowest edge of the prism which is in the same vertical plane as the axis. Find the least velocity with which the particle must be projected at right angles to this plane so that the string, which is always kept tight, may be drawn completely round the prism.

38. Two masses, $m$ and $m'$, are connected by an inextensible string of length $a$. The extremity $A$, to which $m$ is attached, is compelled to move with uniform acceleration $f$ in a straight line under a force $P$ in that line, and the extremity $B$, to which $m'$ is attached, is compelled to describe a circle round $A$ with uniform angular velocity $\omega$ under the action of a force $Q$ perpendicular to $AB$. Find $P$ and $Q$, and shew that the least value of $P$ is $mf - \dfrac{m'a^2\omega^4}{4f}$, provided that $a\omega^2$ is $< 2f$.

39. Two particles of equal mass are joined by a rod without weight and are placed in a horizontal position on a plane inclined to the horizon at an angle $a$; the plane is smooth with the exception of a rough spot on which one of the particles, $A$, rests; the coefficient of friction being $\rho \tan a$, shew that, if $\rho$ be $< 4$, the particle $A$ will begin to move when the rod has turned through an angle $\sin^{-1}\left[\dfrac{\rho^2-1}{15}\right]^{\frac{1}{2}}$.

What happens if $\rho$ be $> 4$

# CHAPTER VIII.

## SIMPLE HARMONIC MOTION. CYCLOIDAL AND PENDULUM MOTIONS.

**177. Theorem.** *If a point* Q *describe a circle with uniform angular velocity and if* P *be the foot of the perpendicular drawn from* Q *upon a fixed diameter* AOA′ *of the circle, to shew that the acceleration of* P *is directed towards the centre,* O, *of the circle and varies as the distance of* P *from* O, *and to find the velocity of* P *and its time of describing any space.*

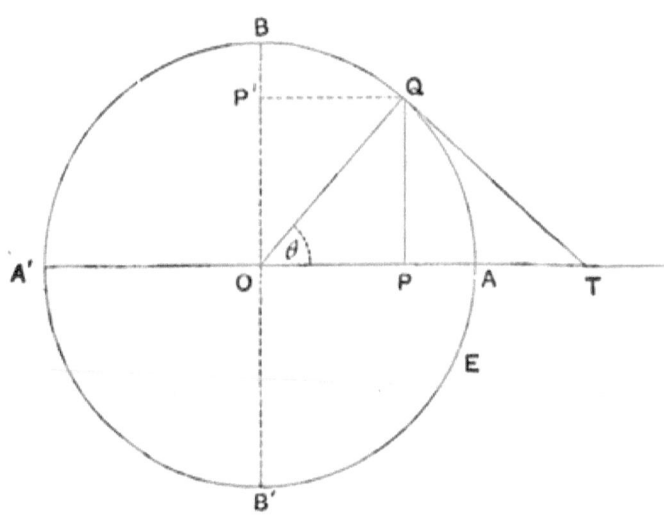

# SIMPLE HARMONIC MOTION.

The velocity and acceleration of $P$ are the same as the resolved parts, parallel to $AO$, of the velocity and acceleration of $Q$.

Let $\omega$ be the constant angular velocity with which the point $Q$ describes the circle.

Let $a$ be the radius of the circle and let the angle $QOA$ be $\theta$. Draw $QT$ the tangent at $Q$.

By Art. 156, Cor. 1, the acceleration of $Q$ is $a\omega^2$ towards $O$.

$\therefore$ the acceleration of $P$ along $PO = a\omega^2 \cos\theta = \omega^2 \cdot OP$, and therefore varies as the distance of $P$ from the centre of the circle.

Also the velocity of $P$
$$= a\omega \cos QTO = a\omega \sin\theta = \omega \cdot PQ = \omega\sqrt{a^2 - x^2},$$
where $OP$ is $x$.

This velocity is zero at $A$ and $A'$, and greatest at $O$.

Also the acceleration vanishes and changes its sign as the point $P$ passes through $O$.

The point $P$ therefore moves from rest at $A$, has its greatest velocity at $O$, comes to rest again at $A'$, and then retraces its path to $A$.

Also the time in which $P$ describes any distance $AP$
= time in which $Q$ describes the arc $AQ$
$$= \frac{\theta}{\omega} = \frac{1}{\omega} \cos^{-1}\left(\frac{x}{a}\right).$$

$\therefore$ time from $A$ to $A' = \dfrac{\pi}{\omega}$.

Also time from $A$ to $A'$ and back again to $A = \dfrac{2\pi}{\omega}$.

**178. Simple Harmonic Motion. Def.** *If a point move in a straight line so that its acceleration is always* *directed towards, and varies as its distance from, a fixed point in the straight line the point is said to move with simple harmonic motion.*

The point $P$ in the previous article moves with simple harmonic motion.

From the results of the previous article we see, by equating $\omega^2$ to $\mu$, that if a point move with simple harmonic motion, starting from rest at a distance $a$ from the fixed centre $O$, and moving with acceleration $\mu . OP$ then

(1) its velocity when at a distance $x$ from $O$ is
$$\sqrt{\mu(a^2 - x^2)},$$

(2) the time that has elapsed when the point is at a distance $x$ from $O$ is $\dfrac{1}{\sqrt{\mu}} \cos^{-1} \dfrac{x}{a}$,

(3) the time that elapses before it is again in its initial position is $\dfrac{2\pi}{\sqrt{\mu}}$.

**179.** The range, $OA$ or $OA'$, of the moving point on either side of centre $O$ is called the **Amplitude** of the motion.

The time that elapses from any instant till the instant in which the moving point is again moving through the same position with the same velocity and direction is called the **Periodic Time** of the motion.

It will be noted that *the periodic time*, $\dfrac{2\pi}{\sqrt{\mu}}$, *is independent of the amplitude of the motion.*

**180.** The **Phase of a** simple harmonic motion is the fraction of the whole period that has elapsed since the point last passed through its extreme position in the positive direction.

**Thus** when the particle is at $P$ the **phase is** $\dfrac{\theta}{2\pi} \cdot T$, where $T$ is the **periodic** time.

If the time be **measured** from the instant at which the corresponding point $Q$ is at $E$ the angle $EOA$ is called the **Epoch and is** generally denoted by $\epsilon$.

If $P$ be **the position of the moving point at time** $t$ **and** $T$ be the periodic time the angle $EOQ$ is $2\pi \dfrac{t}{T}$, and hence the distance $OP$ is $a \cos\left(\dfrac{2\pi}{T} t - \epsilon\right)$ or $a \cos(\sqrt{\mu} t - \epsilon)$.

**181.** If $BOB'$, Fig. Art. 177, be **the** diameter perpendicular **to** $AOA'$ **and** $QP'$ **be drawn** perpendicular to it, the point $P'$ will have **a simple harmonic motion whose amplitude and period are the same as those of** $P$ **but whose** phase is less than **that of** $P$ by one-quarter **of a period**; for when $Q$ is at $B$ the phases are respectively zero and $\tfrac{1}{4}T$.

Hence **we see that a point** which possesses two rectangular simple **harmonic motions of the** same period and amplitude, but of phases differing by one-quarter **of a period, moves in a circle whose** radius is the common amplitude.

**182.** Two simple harmonic motions **in the same** straight line, of the same period, **give** a harmonic motion **of the same period.** For the distance of the point **must** $= a \cos(\sqrt{\mu} t - \epsilon) + a' \cos(\sqrt{\mu} t - \epsilon')$

$$= a'' \cos(\sqrt{\mu} t - \epsilon''),$$

where $a'' \cos \epsilon'' = a \cos \epsilon + a' \cos \epsilon'$, **and** $a'' \sin \epsilon'' = a \sin \epsilon + a' \sin \epsilon'$.

Hence $a'' = \sqrt{a^2 + a'^2 + 2aa' \cos(\epsilon - \epsilon')}$ and $\tan \epsilon'' = \dfrac{a \sin \epsilon + a' \sin \epsilon'}{a \cos \epsilon + a' \cos \epsilon'}$,

so that the new **amplitude** is the diagonal of a parallelogram whose sides are the **original** amplitudes inclined to one another at an angle equal to **the difference of** the epochs.

**183.** The time **of vibration in simple** harmonic motion **may, by** consideration of dimensions **only,** be shewn to be independent of the extent **of** oscillation. For, since the acceleration at any point is $\mu \times$ a distance, the dimensions of $\mu$ must be $-2$ **in time.** Now the time of a complete oscillation can only depend on $\mu$ and $a$. Let it equal $\mu^x \cdot a^y$ so that its dimensions are $[T]^{-2x}[L]^y$. Hence $x = -\frac{1}{2}$ and $y = 0$, so that the required time varies as $\dfrac{1}{\sqrt{\mu}}$ and is independent of $a$.

**184.** *Extension to motion in a curve.*

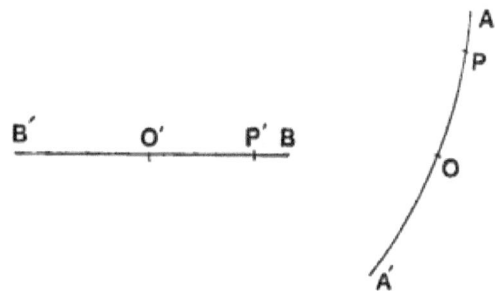

Suppose that a moving point $P$ is describing a portion, $AOA'$, of a curve of any shape starting from rest at $A$ and moving so that its tangential acceleration is always along the arc towards $O$ and equal to $\mu \cdot \text{arc } OP$, then the preceding proposition is true with slight modifications.

For let $O'B$ be a straight line equal in length to the arc $OA$ and let $P'$ be a point describing it with acceleration $\mu \cdot O'P'$; also let $O'P' = \text{arc } OP$.

Since the acceleration of $P'$ in its path is always the same as that of $P$, the velocities acquired in the same time are the same and the times of describing the same distances are the same.

Hence

(1) The velocity of $P$

$$= \sqrt{\mu(O'B^2 - O'P'^2)} = \sqrt{\mu \{(\text{arc } OA)^2 - (\text{arc } OP)^2\}},$$

(2) Time from $A$ to $P$

$$= \frac{1}{\sqrt{\mu}} \cos^{-1}\left(\frac{O'P'}{O'B}\right) = \frac{1}{\sqrt{\mu}} \cos^{-1}\left(\frac{\text{arc } OP}{\text{arc } OA}\right),$$

(3) Time from $A$ to $A'$ and back again $= \dfrac{2\pi}{\sqrt{\mu}}$.

### EXAMPLES.

1. A point, moving with simple harmonic motion, has a velocity of 4 feet per second when passing through the centre of its path and its period is $\pi$ seconds; what is its velocity when it has described one foot from the position in which its velocity is zero?

*Ans.* $2\sqrt{3}$ feet per second.

2. A mass of one gramme vibrates through a millimetre on each side of the middle point of its path 256 times per second; assuming its motion to be simple harmonic, shew that the maximum force upon the particle is $\frac{1}{10}(512\pi)^2$ dynes.

3. A horizontal shelf moves vertically with simple harmonic motion whose complete period is one second; find the greatest amplitude in centimetres that it can have so that objects resting on the shelf may always remain in contact with it.

*Ans.* Nearly 25 centimetres.

4. A particle is oscillating in a straight line about a centre of force, $O$, towards which when at distance $r$ the force is $n^2 r$, and $a$ is the amplitude of the oscillation; when at a distance $a\dfrac{\sqrt{3}}{2}$ from $O$ the particle receives a blow in the direction of motion which generates a velocity $na$. If this velocity be away from $O$, shew that the new amplitude is $a\sqrt{3}$.

5. A particle is initially at a distance $c$ from a fixed point and its acceleration is directed towards, and is equal to $\mu$ times its distance from the fixed point; shew that it will arrive at the fixed point in time

$\dfrac{1}{\sqrt{\mu}} \sin^{-1} \dfrac{c\sqrt{\mu}}{\sqrt{V^2 + \mu c^2}}$, where $V$ is the velocity of projection.

6. An elastic string is extended between two fixed points to twice its natural length and a particle, of mass $m$, is fastened to the middle point of the string. The particle is drawn towards one fixed point through a distance equal to half its distance from it and then let **go**. Find the greatest velocity subsequently attained and, **if** $a$ be the natural length of the string, shew that the time of an oscillation is $\pi \sqrt{\dfrac{ma}{\lambda}}$, where $\lambda$ is the modulus of elasticity of the string.

7. **One end of an** elastic string, whose modulus of elasticity is $\lambda$, is fixed to a point on a smooth horizontal table and the other end is tied to a particle, of mass $m$, which is lying **on** the table. The particle is pulled to a distance from the point of attachment of the string equal to twice the natural length, $a$, of the string and then **let** go; shew that the time **of** a complete oscillation is $2(\pi+2)\sqrt{\dfrac{am}{\lambda}}$.

8. An elastic **string without weight,** of which the unstretched length is $l$ and the modulus of elasticity the weight of $n$ **ozs., is** suspended by one end and a mass of $m$ ozs. is attached to the **other;** shew that the time of a vertical oscillation **of the mass is** $2\pi\sqrt{\dfrac{m}{n}\dfrac{l}{g}}$.

9. Find the resultant of two simple **harmonic vibrations in** the same direction and of equal periodic time, the amplitude **of one** being twice that of the other and its phase a quarter of a period in advance.

*Ans.* A simple harmonic vibration of amplitude $\sqrt{5}$ times that of **the** first and whose phase is in advance of the first by $(\tan^{-1} 2)/2\pi$ **of a period.**

## *Motion on a Cycloid and Pendulums.*

185. In the following **articles** we **shall** discuss the motion of simple forms of Pendulums. Intimately connected with this motion however is motion down a smooth wire in the form of a curve called a Cycloid. It will be desirable to define and call attention to its properties.

**186. Cycloid. Def.** *If a circle roll, without sliding, in one plane on a fixed straight line the path of any point fixed on the circumference of the circle and moving with it is called a Cycloid.*

[It is the path of a speck of mud on the edge of a cartwheel which is rolling in a straight line on a level road.]

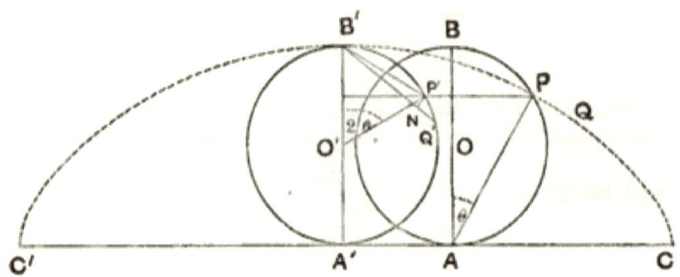

Thus let the circle $APB$ roll on the straight line $C'A'C$ so that a point $P$ fixed on its circumference describes the arc $C'B'PC$. This curve is a portion of a cycloid.

Let $A'P'B'$ be the position of the circle when the fixed point is at its greatest distance from $CC'$.

Then $B'$ is called the vertex, $A'B'$ the axis, and $C'C$ the base of the cycloid.

If a perpendicular from $P$ on $A'B'$ meet the circle $A'P'B'$ in $P'$ then

(1) The tangent to the cycloid at $P$ is parallel to $B'P'$,

(2) The arc $B'P$ = twice the line $B'P'$, and

(3) The radius of curvature at $P$ is equal to twice the line $AP$, i.e., is equal to twice the portion of the normal intercepted between the curve and the base.

In the following articles we shall give proofs of these propositions. For other proofs the student may refer to Thomson and Tait's *Natural Philosophy*, Vol. I., Art. 93, or to Edwards' *Differential Calculus*, Art. 356 etc.

187. Let $Q$ be a consecutive point on the cycloid and let perpendiculars from $P$ and $Q$ upon $A'O'B'$ meet the circle $A'P'B'$ in $P'$ and $Q'$. Also draw $P'N$ perpendicular to $B'Q'$.

We have shewn (Art. 40, Ex. 1) that the point $P$ is for the instant turning about the point $A$ with the same angular velocity with which the circle is turning about its centre.

Hence the point $P$ is moving perpendicular to $AP$ and the tangent at $P$ is therefore $BP$ which is parallel to $B'P'$.

Again since $P$ has the same angular velocity about $A$ that the circle has about its centre and since, whilst the circle turns through a small angle $P'O'Q'$, $P$ advances to $Q$,

$$\therefore \frac{PQ}{P'Q'} = \frac{AP}{O'P'} = \frac{2a\cos\theta}{a} = 2\cos B'A'P' = 2\cos P'Q'N = 2\frac{NQ'}{P'Q'}.$$

$$\therefore PQ = 2 \cdot NQ'.$$

Now $P'B'Q'$ is very small and $B'P'N$, $B'NP'$ are ultimately right angles so that $B'N = B'P'$ ultimately.

$$\therefore PQ = 2NQ' = 2(B'Q' - B'N) = 2(B'Q' - B'P').$$

Hence the increase in the arc $B'P$ is equal to twice the increase in the line $B'P'$; also the arc $B'P$ and the line $B'P'$ vanish together at the vertex; hence the arc $B'P$ is twice the line $B'P'$.

It follows that the whole arc $B'C$ is twice $B'A'$ or four times the radius of the rolling circle.

188. *Radius of curvature.* Since $B'P'$, $B'Q'$ are parallel to the tangents at $P$ and $Q$, the angle between these tangents is $P'B'Q'$ or $\frac{1}{2}P'O'Q'$.

Hence the radius of curvature at $P$

$= $ Lt. arc $PQ \div$ angle between the tangents at $P$ and $Q$

$= $ Lt. $2$ arc $P'Q' \cos\theta \div \frac{1}{2} \angle P'O'Q'$

$= 4\cos\theta \times $ Lt. arc $P'Q' \div \angle P'O'Q'$

$= 4a\cos\theta = 2AP.$

**189.** Let $C'B'C$ be a cycloid with axis vertical and vertex downward.

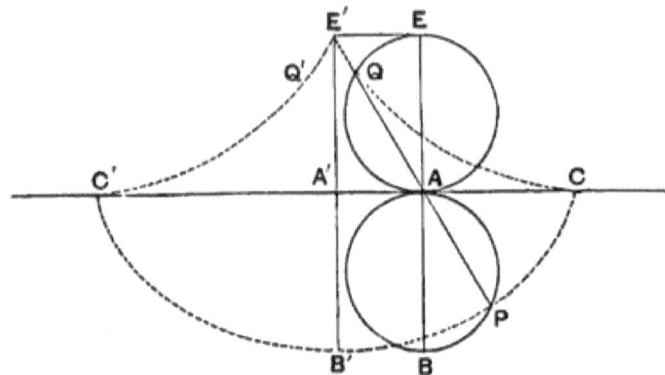

Draw the generating circle $APB$ in the position in which it intersects the cycloid in $P$, so that $AP$ is the normal at $P$.

Produce $B'A'$ to $E'$ making $A'E'$ equal to $A'B'$ and draw $E'E$ parallel to $A'A$ to meet $BA$ produced in $E$; also produce $PA$ to $Q$ making $AQ$ equal to $PA$.

Then since the triangles $BAP$, $EAQ$ are equal in all respects, $EQA$ is a right angle and a circle on $EA$ as diameter will pass through $Q$, so that

$$\text{arc } EQ = \text{arc } BP = \text{line } A'A = E'E.$$

Hence if $Q$ be a fixed point on the circle $EQA$, and if the circle, starting with $Q$ in contact with $E'E$ at $E'$, roll along $E'E$ the fixed point will trace out a portion of a cycloid, $E'QC$, having its cusp at $E'$ and vertex at $C$.

Also the $\qquad$ arc $QC = 2$ . line $AQ = PQ$.

Hence it follows that if a string be wrapped tightly round the curve $E'QC$, one end being fixed at $E'$ and the other being initially at $C$, this end will when the string is unwrapped describe the arc $CPB'$; if the string then wrap itself upon a similar portion of a cycloid, having vertex at $C'$ and cusp at $E'$, its end will describe the other portion $B'C'$ of the original cycloid.

The student who is acquainted with the Differential Calculus will observe that $Q$ is the centre of curvature of the curve at $P$ and that the arc $E'QC$ is therefore a portion of the evolute of the original curve, so that the above proposition is a particular case of the proposition which is true for all curves and their evolutes.

## CYCLOID UNDER GRAVITY.

**190. Theorem.** *If a smooth cycloid be placed in a vertical plane, with its axis vertical and its vertex downward, to shew that the time of oscillation of a particle from rest to rest is the same whatever be the point of the curve from which the particle starts.*

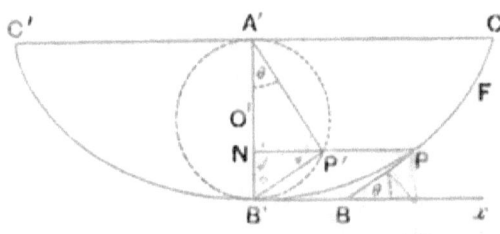

Let $C'B'C$ be the cycloid, $CC'$ being the base and $B'$ the vertex, $P$ any point **on the arc,** and $A'P'B'$ the circle on $A'B'$ as diameter.

Draw $PN$ perpendicular to $A'B'$ meeting the circle in $P'$ and let the tangent at $P$ meet the tangent at $B'$ in $B$.

**Then,** as in Art. **186**, $PB$ is parallel to $P'B'$ and the arc $B'P$ is twice the line $B'P'$.

Hence $\angle PBx = \angle P'B'x = \angle B'A'P' = \theta$.

Let a particle start from rest at a point $F$ of the arc and slide down the arc and let $P$ be its position at any time.

Its tangential acceleration $= g \sin \theta = g \sin B'A'P'$
$= g \cdot \dfrac{B'P'}{2a} = \dfrac{g}{4a} \cdot \text{arc } B'P$, where $A'B'$ is $2a$.

Hence the particle moves so that its tangential acceleration is **always** along the arc toward $B'$ and equal to $\dfrac{g}{4a} \cdot \text{arc } B'P$.

Hence, by Art. 184, the time of a complete oscillation is $2\pi\sqrt{\dfrac{4a}{g}}$ and is therefore independent of the position of the point $F$ from which the particle started.

Hence the time of oscillation in an inverted cycloid is independent of the extent of the oscillation.

On account of this property the cycloid is called the isochronous curve under gravity.

**191.** The property proved in the previous article will be still true if, instead of the material curve, we substitute a string tied to the particle in such a way that the particle describes a cycloid and the string is always normal to the curve.

This, by Art. 189, may be done by attaching the string at $E'$ and compelling it to wrap and unwrap itself alternately upon two metal cheeks in the shape of the arcs $E'QC$ and $E'Q'C'$; in actual practice a pendulum is only required to swing through a small angle so that only very small portions of the two arcs near $E'$ are required.

**192. Simple Pendulum.** A particle tied to a string, one end of which is fixed, and swinging in a vertical plane is called a simple pendulum.

Referring to Fig. Art. 189, since the arc $E'C$ is equal to $4a$, a pendulum of length $l$ describes a cycloid the radius of whose generating circle is $a$, where $l = 4a$.

Hence the time of a complete oscillation of the pendulum = periodic time in the cycloid

$$= 2\pi\sqrt{\dfrac{4a}{g}} = 2\pi\sqrt{\dfrac{l}{g}}.$$

**193.** If, instead of compelling the particle in the simple pendulum to describe a cycloid, we **allow it** to describe a circle about the other end as centre the time of oscillation is still **very** approximately constant, provided the angle through which the string oscillates be small. This can be shewn as follows.

**Theorem.** *If a particle be tied by a string to a fixed point and allowed to oscillate through a small angle about the vertical position, to shew that the time of a complete oscillation is* $2\pi\sqrt{\dfrac{l}{g}}$, *where* l *is the length of the* **string**.

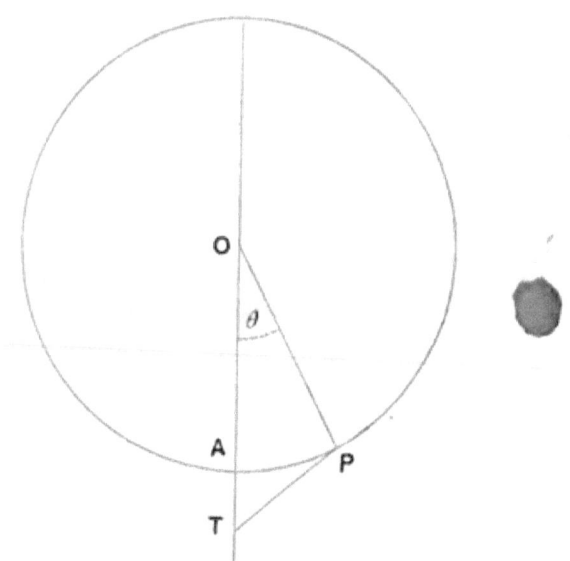

Let $O$ be the fixed point, $OA$ a vertical line, $AP$ a portion of the arc described by the particle, and let the angle $AOP$ be $\theta$.

If $PT$ be the tangent at $P$ meeting $OA$ in $T$, the acceleration of the bob along $PT$

L. D.

$= g \sin \theta$

$= g\theta$ (approximately, if $\theta$ be small)

$= \dfrac{g}{l} \times \text{arc } AP$.

∴ as before, the time of a double oscillation is independent of the extent of the oscillation and is equal to $2\pi \sqrt{\dfrac{l}{g}}$.

194. The above result although not mathematically accurate is very approximately so. If we take into consideration the angle $\alpha$ through which the pendulum swings on each side of the vertical and neglect powers of $\sin \alpha$ above the second, the time of oscillation is

$$2\pi \sqrt{\dfrac{l}{g}} \left[ 1 + \dfrac{1}{4} \sin^2 \dfrac{\alpha}{2} \right].$$

[For a proof the student may refer to Greenhill's *Solutions of Cambridge Problems and Riders*, 1875.]

If $\alpha$ be 5°, the result of the preceding article is within one-two thousandth part of the accurate result, so that a pendulum which beats seconds for very small oscillations would lose about 40 seconds per day if made to vibrate through 5° on each side of the vertical.

195. *Length of a seconds pendulum.* A seconds pendulum is one which makes half a complete oscillation (viz. from $C$ to $C'$ in Fig. Art. 189) in one second.

Hence, if $l$ be its length, we have $1 = \pi \sqrt{\dfrac{l}{g}}$.

$$\therefore l = \dfrac{g}{\pi^2}.$$

For an approximate value, putting $g = 32\cdot2$ and $\pi = \frac{22}{7}$, we have $l = 39\cdot12$ inches.

If we use the centimetre-second system we have $l = 99\cdot3$ centimetres.

For the latitude of London more accurate values are $39\cdot13929\ldots$ inches and $99\cdot413\ldots$ centimetres.

**196.** The simple pendulum of which we have spoken is idealistic. In practice, a pendulum consists of a wire whose mass, although small, is not zero and a bob at the end which is not a particle. Whatever be the shape of the pendulum, the simple pendulum which oscillates in the same time as itself is its **simple equivalent pendulum**.

The discussion of the connection between a rigid body and its simple equivalent pendulum is not within the range of this book. We may, however, mention that a uniform rod, of small section, swings about one end in the same time as a simple pendulum of two-thirds its length, whilst the length of the simple equivalent pendulum of a circular lamina swinging about an axis through the end of a diameter perpendicular to its plane is three-quarters of the diameter of the lamina.

**197.** *Acceleration due to gravity.* Newton shewed as a fundamental law of nature that every particle attracts every other particle with a force which varies directly as the product of the masses and inversely as the square of the distance between them. From this fact it can be shewn, as in any treatise dealing with Attractions, that a sphere attracts any particle *outside* itself just as if the whole mass of the sphere were collected at its centre, and hence that the acceleration caused by its attraction varies inversely as the square of the distance of the particle from its centre; similarly the attraction on a particle *inside* the earth can be shewn to vary directly as its distance from the centre of the earth.

Hence if $g_1$ be the value of gravity at a height $h$ above the earth's surface, $g$ the value at the surface, and $r$ the earth's radius, then $g_1 : g :: \dfrac{1}{(r+h)^2} : \dfrac{1}{r^2}$,

$$\text{so that } g_1 = g\left(\dfrac{r}{r+h}\right)^2.$$

So if $g_2$ be the value at the bottom of a mine, of depth $d$, we have $g_2 = g\dfrac{r-d}{r}$.

198. We will now investigate the effect on the time of oscillation of a simple pendulum due to a *small* change in the value of $g$, and also the effect due to a *small* change in its length.

*If a pendulum, of length* l, *make* n *complete oscillations in a given time, to shew that*

(1) *If* g *be changed to* g+G, *the number of oscillations gained is* $\dfrac{n}{2}\dfrac{G}{g}$,

(2) *If the pendulum be taken to a height* h *above the earth's surface the number of oscillations lost is* $n\dfrac{h}{r}$, *where* r *is the radius of the earth,*

(3) *If it be taken to the bottom of a mine of depth* d, *the number lost is* $\dfrac{n}{2}\dfrac{d}{r}$,

(4) *If its length be changed to* l+L, *the number lost is* $\dfrac{n}{2}\dfrac{L}{l}$.

Let $T$ be the original time of oscillation, $T'$ the new time of oscillation, and $n'$ the new number of oscillations in the given time so that

$$nT = n'T'.$$

## OSCILLATION OF A PENDULUM.

(1) In this case $T = 2\pi \sqrt{l/g}$; $T' = 2\pi \sqrt{l/(g+G)}$.

Hence $\dfrac{n'}{n} = \dfrac{T}{T'} = \sqrt{1 + \dfrac{G}{g}} = 1 + \dfrac{1}{2}\dfrac{G}{g}$, approximately,

(using the Binomial Theorem and neglecting squares of $\dfrac{G}{g}$).

Hence the number of oscillations gained $= n' - n = \dfrac{n}{2}\dfrac{G}{g}$.

So if $g$ become $g - G$, the number lost is $\dfrac{n}{2}\dfrac{G}{g}$.

(2) If $g - G$ be the value of gravity at a height $h$ we have

$$\dfrac{g-G}{g} = \dfrac{r^2}{(r+h)^2} = \left(1 + \dfrac{h}{r}\right)^{-2} = 1 - \dfrac{2h}{r} \text{ approximately.}$$

$\therefore G = g\,\dfrac{2h}{r}$ and hence, as in (1), **the number** of oscillations lost is $n\dfrac{h}{r}$.

(3) If $g - G$ be the value at **a depth** $d$ we have $g - G : g :: r - d : r$, so that the number **of** oscillations lost $= \dfrac{1}{2}\dfrac{G}{g} = \dfrac{n}{2}\dfrac{d}{r}$.

(4) when the length $l$ **of** the pendulum is changed to $l + L$ we have

$$T = 2\pi \sqrt{l/g}; \quad T' = 2\pi \sqrt{(l+L)/g}.$$

$\therefore \dfrac{n'}{n} = \dfrac{T}{T'} = \left(1 + \dfrac{L}{l}\right)^{-\frac{1}{2}} = 1 - \dfrac{1}{2}\dfrac{L}{l}$ approximately.

Hence the number of oscillations lost $= n - n' = \dfrac{n}{2}\dfrac{L}{l}$.

**199.** From the preceding article it follows that the height of a mountain or the depth of a mine could be found by finding the number of oscillations lost by a pendulum which beats seconds on the surface of the earth.

If the point above the sea level at which the pendulum beats be on a considerable table land the effect of the **attraction of the portion** of the earth which **is** between the pendulum and sea level could be allowed for by taking the value of gravity there as $g\left(1 - \frac{5h}{4r}\right)$ instead of $g\left(1 + \frac{h}{r}\right)^{-2}$ and then the number of oscillations lost would be $\frac{5n}{8}\frac{h}{r}$ instead of $n\frac{h}{r}$.

200. The value of $g$ has been most accurately found by means of timing the oscillation of a pendulum. This may be done as follows.

The pendulum on which we experiment is set swinging in front of a clock whose pendulum we know to be beating true seconds, and they are started so that they are both vertical at the same instant and the time is carefully noted at which they are again both vertical together and moving in the same direction; we then know that the pendulum has gained (or lost) one complete oscillation on the seconds pendulum. Suppose, for example, that they coincide at the end of 300 seconds, then in that time the seconds pendulum has made 150 complete oscillations and the other pendulum 149 so that the time of its complete oscillation would be $\frac{300}{149}$ seconds.

The time of an oscillation may be found much more accurately by finding the time that would elapse before the experimental pendulum had lost a large number, say 20, oscillations and then taking the average.

The length of the simple equivalent pendulum corresponding to our experimental pendulum may be then measured and the value of $g$ determined by means of the formula $T = 2\pi\sqrt{\frac{l}{g}}$.

# EXAMPLES ON CHAPTER VIII.

**201. *Verification of the law of gravity by means of the moon's motion.*** We may show roughly the truth of the law of gravitation by finding the time that the moon would take to travel round the earth on the assumption that it is kept in its orbit by means of the earth's attraction.

Let $f$ be the acceleration of the moon due to the earth's attraction; then, since the distance between the centres of the two bodies is roughly 60 times the earth's radius, we have $f : g :: \dfrac{1}{(60r)^2} : \dfrac{1}{r^2}$, so that $f = g/3600$.

Let $v$ be the velocity of the moon round the earth so that

$$v^2/60r = f = g/3600.$$

$$\therefore v^2 = \frac{gr}{60}.$$

$\therefore$ the periodic time of the moon $= 2\pi \times 60r \div v = 2\pi \cdot 60 \times \sqrt{\dfrac{60r}{g}}$ seconds.

Taking the radius of the earth to be 4000 miles and $g$ as 32·2 this time is 27·4 days, which is approximately the observed time of revolution.

## EXAMPLES. CHAPTER VIII.

1. How many oscillations will a pendulum of length 53·41 centimetres make in 242 seconds at a place where $g = 981$?

2. Shew that a pendulum one mile in length will oscillate in $\frac{1}{7}\sqrt{22}$ minutes nearly.

3. A pendulum 37·8 inches long makes 183 beats in three minutes at a certain place; find the acceleration due to gravity there.

4. A pendulum oscillating seconds at one place is carried to another place where it loses two seconds per day; compare the accelerations due to gravity at the two places.

# EXAMPLES ON CHAPTER VIII.

5. A clock which at the surface of the earth gains 10 seconds a day loses 10 seconds a day when taken down a mine; compare the accelerations due to gravity at the top and bottom of the mine and find the depth of the mine.

6. If a seconds pendulum be carried to the top of a mountain half a mile high how many seconds will it lose per day, assuming the earth's centre to be 4000 miles from the foot of the mountain, and by how much must it be shortened so that it may beat seconds at the summit of the mountain?

7. Shew that the height of a hill at the summit of which a seconds pendulum loses $n$ beats in 24 hours is approximately $245 \cdot n$ feet.

8. If a pendulum oscillating seconds be lengthened by its hundredth part, find the number of oscillations it will lose in a day.

9. A simple pendulum performs 21 complete vibrations in 44 seconds; on shortening its length by 47·6875 centimetres it performs 21 complete vibrations in 33 seconds; find the value of $g$.

10. A clock with a seconds pendulum loses 9 seconds per day; find roughly the required alteration in the length of the pendulum.

11. If $l_1$ be the length of an imperfectly adjusted seconds pendulum which gains $n$ seconds in an hour and $l_2$ the length of one which loses $n$ seconds at the same place; shew that the true length of the seconds pendulum is the harmonic mean between $l_1$ and $l_2$.

12. A balloon ascends with a constant acceleration and reaches a height of 900 feet in one minute. Shew that a pendulum clock carried in the balloon will gain at the rate of 27·8 seconds per hour.

13. Two clocks identical in all respects are placed at opposite points on the earth's surface, at one of which the moon is vertically overhead. Assuming the earth and moon to remain at rest and that the mass of the earth is 80 times that of the moon, shew that if the clocks be started together they will at the end of twenty-four hours differ by about $\frac{3}{10}$ths of a second.

14. The bob of a pendulum which is hung close to the face of a vertical cliff is attracted by the cliff with a force which would produce an acceleration $f$ in the bob; shew that the time of a complete oscillation is $2\pi\sqrt[4]{l^2/(g^2+f^2)}$, where $l$ is the length of the pendulum.

15. Prove that a seconds pendulum, if carried to the moon, would oscillate in $1\frac{3}{4}$ seconds, assuming the mass of the earth to be 49 times that of the moon and the radius of the earth 4 times that of the moon.

16. A railway train is moving uniformly along a curve at the rate of 60 miles per hour and in one of the carriages a seconds pendulum is observed to beat 121 times in two minutes. Prove that the radius of the curve is about a quarter of a mile.

17. Assuming the earth to be a sphere whose attraction per unit of mass is the same at all points of its surface, prove that a pendulum which beats seconds at the poles will lose $30\,m\cos^2\lambda$ beats per minute in latitude $\lambda$, where the ratio of the weight of the body at the poles to its weight at the equator is $1+m$ to $1$.

18. If a seconds pendulum be formed of a heavy particle suspended by a string of length $l$ from a point $A$ in a vertical wall, and if a nail jut out from the wall at a distance $\dfrac{l}{n}$ below $A$ so as to catch the string once in each complete oscillation find the time of oscillation, $n$ being large.

19. In cycloidal motion the vertical velocity of the particle is greatest when it has described half its vertical descent.

20. A particle of mass $m$ falls down a cycloid under gravity starting from the cusp; shew that the pressure on the cycloid at any point is $2mg \cos \psi$, where $\psi$ is the inclination to the horizon of the tangent to the cycloid at that point.

21. When a particle falls from rest down the arc of a cycloid prove that it describes half the path to the lowest point in two-thirds the time to the lowest point.

22. If a particle slide down a smooth cycloid, whose axis is vertical and vertex downwards, from a point whose arcual distance from the vertex is $b$, prove that its velocity at time $t$ from the start is $\dfrac{2\pi b}{\tau} \sin \dfrac{2\pi t}{\tau}$, where $\tau$ is the time of a complete oscillation of the particle.

23. A particle is just displaced from rest at the vertex of a smooth cycloid, whose axis is vertical and vertex upwards, and which rests on a horizontal plane; shew that it will leave the cycloid when its direction of motion makes an angle of 45° with the vertical and that it will strike the plane at a time $(\sqrt{3}-1)\sqrt{\dfrac{a}{g}}$ after it leaves the cycloid, $a$ being the radius of the generating circle of the cycloid.

24. Shew that the time a train, if unresisted, takes to pass through a tunnel under a river in the form of an arc of an inverted cycloid of length $2s$ and height $h$ cut off by a horizontal line is

$$\frac{s}{\sqrt{2gh}} \cos^{-1}\left(\frac{v^2 - 2gh}{v^2 + 2gh}\right),$$

where $v$ is the velocity with which the train enters and leaves the tunnel.

25. A particle moves in a circle drawn on the deck of a ship with a uniform speed equal to that of the ship. Find the path it describes in space. Also find the path of the particle if the given uniform speed be less than that of the ship.

26. A pendulum, of length $l$, has one end of its string fastened to a peg on a smooth plane inclined at an angle $a$ to the horizon; with the string and the weight on the plane its time of oscillation is two seconds. Find $a$ having given that a pendulum of length $\dfrac{l}{2\sqrt{2}}$ oscillates in one second when suspended vertically.

27. In a conical pendulum shew that the time of revolution is ultimately $2\pi\sqrt{\dfrac{l}{g}}$ when the inclination is very small.

28. If the length $l$ of a simple pendulum be comparable with the radius $a$ of the earth, shew that its time of oscillation is $\pi\sqrt{la/g(a-l)}$.

29. A point moves in a path produced by the combination of two simple harmonic motions of equal amplitude in two rectilinear directions, the periods of the components being as $1:3$; find the paths described when the epochs are the same and when they differ by 90°.

30. If a frictionless airless tunnel were made from London to Paris, prove that a train would traverse the tunnel under the action of gravity in less than three-quarters of an hour.

Shew that the same theorem would be true whatever two points were taken on the earth's surface.

# CHAPTER IX.

## MOTION OF A PARTICLE ABOUT A FIXED CENTRE OF FORCE.

202. **Moment of a velocity. Def.** *The moment of the velocity of a moving point about a fixed point is the product of the velocity of the moving point and the perpendicular drawn from the fixed point upon the direction of motion of the moving point.*

If $O$ be the fixed point and $PA$ represent in magnitude and direction the velocity of the moving point at a

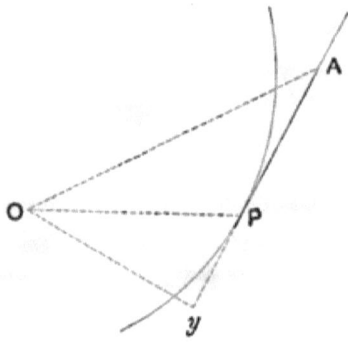

point $P$ of its path, the moment of its velocity about $O$ is

MOMENT OF A VELOCITY. 301

represented by $PA \cdot Oy$, where $Oy$ is the perpendicular upon $PA$. Hence this moment may be represented geometrically by twice the area of the triangle $OPA$.

203. **Theorem.** *If a point describe a curve in one plane, and its acceleration be always directed towards a fixed point in the plane, the moment of its velocity about the fixed point is unaltered.*

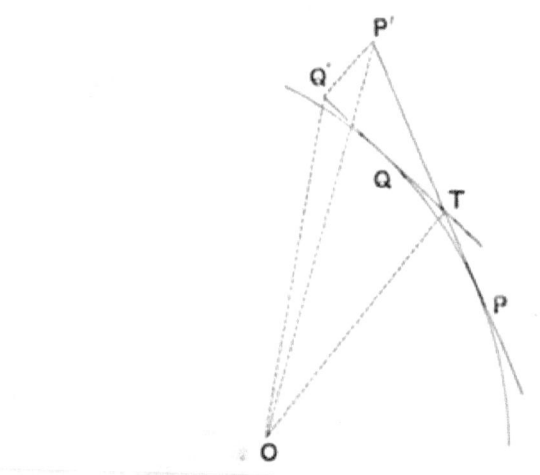

Let $PQ$ be a portion of the path of the moving point, and let the tangents at $P$ and $Q$ meet in $T$. Let $TP'$, $TQ'$ represent the velocities at $P$ and $Q$.

Let $O$ be the fixed point; join $OT$ and $P'Q'$. Since, whilst the point describes the arc $PQ$, its velocity is changed from $TP'$ to $TQ'$, the total change of velocity is represented by some line passing through $T$. But, since the acceleration of the point is always directed towards $O$, the direction of this change of velocity must pass through $O$. Hence the change of velocity whilst the moving point describes the arc $PQ$ is in the direction $TO$.

But, as in Art. 24, this change of velocity is represented by $P'Q'$.

$\therefore P'Q'$ is parallel to $TO$.

Now the moments of the velocities at $P$ and $Q$ about $O$ are represented by twice the areas of the triangles $OTP'$, $OTQ'$ respectively.

Also, since $TO$ and $P'Q'$ are parallel, these two triangles are equal.

Hence the moments of the velocities at $P$ and $Q$ about $O$ are the same.

This is true whatever points $P$ and $Q$ are taken on the path.

Hence the theorem is proved.

*Cor.* By Art. 40, the areal velocity of the moving point about $O$ is $\frac{1}{2} v \cdot p$ and it is therefore constant.

Hence the area traced out by the line joining the moving point to $O$ is proportional to the time of describing the area, i.e. **Equal areas are described in equal times.** This is the form in which Newton enunciates the proposition in Book I., Section II., Prop. I. of the *Principia*.

**204. Theorem.** *A particle is projected from a given point, with any velocity and direction, and moves with an acceleration which is always directed towards a fixed point and varies as the distance from the fixed point; to shew that its path is an ellipse whose centre is at the fixed point, and to determine the other circumstances of the motion.*

Let $C$ be the fixed point towards which the accelera-

## AN ACCELERATION DIRECTED TOWARDS THE CENTRE.

tion is always directed, $A$ the point of projection, and $V$ the velocity of projection.

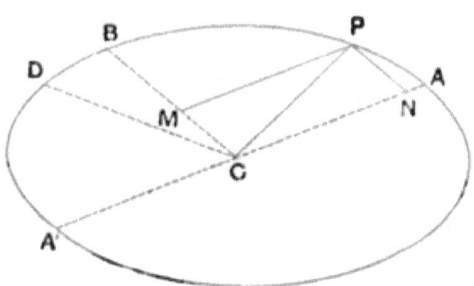

Draw $CB$ parallel **to the direction** of projection. Let $P$ be the position of **the particle** at the end of time $t$ and draw $PN$, $PM$ parallel **to** $CB$ and $CA$ to meet $CA$ and $CB$ in $N$, $M$ respectively. Let the **acceleration towards** $C$ be $\mu \cdot PC$. By **the triangle** of accelerations this acceleration is equivalent to accelerations represented by $\mu \cdot PN$ and $\mu \cdot NC$, i.e. by $\mu \cdot MC$ and $\mu \cdot NC$.

**Hence** the points $N$, $M$ move in the lines $AC, CB$ with simple harmonic motion.

Make $CB$ equal to $V^2/\mu$; then, by Art. 178 (1), $CB$ is the amplitude of the harmonic motion of the point $M$.

Also $CA$ is the amplitude of the motion of the point $N$.

Let $CA$, $CB$ be $a'$ and $b'$ respectively.

**By Art.** 178, (2) we have $CN = a' \cos \sqrt{\mu} t$.

**Also** since $t$ is the **time from** $C$ **to** $M$ and since $\dfrac{\pi}{2\sqrt{\mu}}$ is the time **from** $C$ to $B$, therefore $\dfrac{\pi}{2\sqrt{\mu}} - t$ is the time from $M$ to $B$, or from $B$ to $M$.

$$\therefore CM = b' \cos\left[\sqrt{\mu}\left\{\frac{\pi}{2\sqrt{\mu}} - t\right\}\right],$$

$$\therefore PN = b' \cos\left(\frac{\pi}{2} - \sqrt{\mu}t\right) = b' \sin \sqrt{\mu}t.$$

Hence $\qquad \dfrac{CN^2}{a'^2} + \dfrac{PN^2}{b'^2} = 1.$

Hence $P$ moves on an ellipse of which $CA$ and $CB$ are a pair of conjugate diameters.

**Periodic Time.** The time of describing the ellipse is equal to the periodic time in the simple harmonic motion, and hence is $\dfrac{2\pi}{\sqrt{\mu}}$.

If $h$ be twice the area described in the unit of time, we have

$$\frac{h}{2} \times \frac{2\pi}{\sqrt{\mu}} = \text{area of the ellipse} = \pi ab,$$

where $2a$, $2b$ are the major and minor axes of the ellipse.

Hence $\qquad h = \sqrt{\mu}\, ab.$

**Velocity at any point.** If $v$ be the velocity of the particle at $P$, and $CF$ be the perpendicular from $C$ upon the tangent at $P$, we have

$$v = \frac{h}{CF} = \frac{\sqrt{\mu}\, ab}{CF}.$$

But if $CD$ be the semi-diameter conjugate to $CP$, we have $\qquad CF \cdot CD = ab.$

$\therefore v = \sqrt{\mu} \cdot CD$, and hence the velocity at any point $P$ varies as the diameter conjugate to $CP$.

## MOTION IN A CONIC SECTION.

**205. Theorem.** *A particle is moving in an ellipse with an acceleration which is always directed towards one of the foci; to shew that the acceleration varies inversely as the square of the focal distance and to find the other circumstances of the motion.*

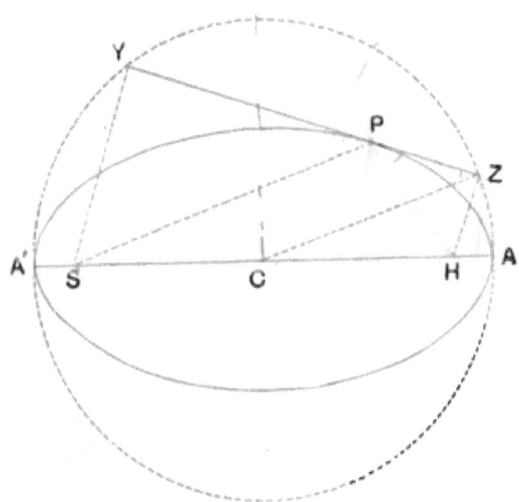

Let $S$ be the focus towards which the acceleration is always directed, $H$ the other focus, and $P$ the position of the particle at any time. Let $v$ be the velocity of the particle at $P$.

Draw $SY$, $HZ$ the perpendiculars upon the tangent at $P$.

Since the acceleration is always towards $S$,
∴ by Art. 203, we have

$$v = \frac{h}{SY} = \frac{h \cdot HZ}{SY \cdot HZ} = \frac{h}{b^2} HZ,$$

if $b$ be the semi-minor axis, and $h$ be twice the area described in unit time.

L. D.

Hence $HZ$ is perpendicular to and proportional to the velocity; therefore the hodograph of the motion is the locus of $Z$ [i.e. the auxiliary circle] turned through a right angle about $H$.

Hence the acceleration in the path is perpendicular and proportional to the velocity of the point $Z$.

Now, since $CZ$ is parallel to $SP$, the angular velocity of $Z$ about $C$ is equal to the angular velocity, $\omega$, of the particle about $S$, and hence the velocity of $Z$ is $a\omega$, where $a$ is the semi-major axis.

∴ the acceleration of $P = \dfrac{h}{b^2} \cdot a\omega$.

But, by Art. 40, $\omega = \dfrac{h}{r^2}$, where $SP$ is $r$.

∴ the acceleration of $P = \dfrac{h^2 a}{b^2} \cdot \dfrac{1}{r^2}$.

Hence the acceleration of the particle is $\dfrac{\mu}{r^2}$, where $\dfrac{h^2 a}{b^2} = \mu$, and therefore

$$h = \sqrt{\mu \times \text{semi-latus-rectum}}.$$

**Periodic Time.** The periodic time

$$= \pi a b \div \tfrac{1}{2} h = 2\pi ab \div \sqrt{\mu \dfrac{b^2}{a}} = \dfrac{2\pi}{\sqrt{\mu}} a^{\frac{3}{2}}.$$

**Velocity at any point.** If $SY$, $HZ$ be $p$, $p'$ respectively, we have $v = \dfrac{h}{p}$.

But the triangles $SPY$, $HPZ$ are similar, and therefore $\dfrac{p}{r} = \dfrac{p'}{r'}$.

$$\therefore \frac{p^2}{r^2} = \frac{pp'}{rr'} = \frac{b^2}{rr'}.$$

$$\therefore \frac{b^2}{p^2} = \frac{r'}{r} = \frac{2a-r}{r}.$$

$$\therefore v^2 = \frac{h^2}{p^2} = \frac{h^2}{b^2} \cdot \frac{2a-r}{r} = \frac{\mu}{a} \cdot \frac{2a-r}{r} = \mu\left(\frac{2}{r} - \frac{1}{a}\right).$$

*Cor.* 1. If in the previous proposition we make $a$ infinite whilst the semi-latus-rectum, $\frac{b^2}{a}$, remains finite, the ellipse becomes a parabola, and the expression for the square of the velocity at any point becomes $\frac{2\mu}{r}$.

*Cor.* 2. If we apply the reasoning of the proposition to the case of a hyperbola we should find that a particle would describe a branch with an acceleration $\frac{\mu}{r^2}$ directed to the focus belonging to the branch; in this case, since in the hyperbola the *difference* of the focal distances of any point is equal to the major axis, the square of the velocity would be $\mu\left(\frac{2}{r} + \frac{1}{a}\right)$.

206. Conversely, if a particle be projected from any point $P$ with any velocity in any direction, and move with an acceleration which is always directed towards a fixed point $S$ and is equal to $\frac{\mu}{(\text{distance})^2}$, it will describe a conic section having $S$ as one of its foci; also the path will be an ellipse, a parabola, or an hyperbola, according as the square of the velocity of projection is less than, equal to, or greater than the quantity $\frac{2\mu}{SP}$.

The orbit may be easily constructed. For the major axis of the orbit is given by $v^2 = \mu \left( \dfrac{2}{SP} - \dfrac{1}{a} \right)$, where $v$ is the velocity of projection. Also the second focus $H$ is such that $SP$, $HP$ make equal angles with the direction of projection and such that $SP + HP$ is equal to the major axis. Hence the position of the major axis is determined.

**207.** The points $A$ and $A'$, Fig. Art. 205, at which the particle is moving in a direction perpendicular to the line joining it to the centre of acceleration, are called the **Apses** of the orbit. The line joining them, which is the major axis of the orbit, is called the **Apse-Line**.

**208. Kepler's Laws.** The astronomer Kepler, by much patient labour, discovered three laws connecting the motions of the various planets about the sun. They are;

1. *Each planet describes an ellipse having the sun in one of its foci.*

2. *The areas described by the radii drawn from the planet to the sun are, in the same orbit, proportional to the times of describing them.*

3. *The squares of the periodic times of the various planets are proportional to the cubes of the major axes of their orbits.*

From the second law we conclude, by Art. 203, that the acceleration of each planet (and therefore the force acting on it) is directed towards the sun.

## EFFECT OF DISTURBING FORCES.

From the **first** law we conclude, by **Art. 205**, that the acceleration of each planet varies inversely as the square of its distance from the sun's centre.

From the third law we conclude, by Art. 205, that the absolute acceleration $\mu$ [that is, the acceleration at unit distance **from the sun**] is the same **for** all the **planets**.

209. *Effect **of** disturbing forces on the path of a particle.* Tangential disturbing force.

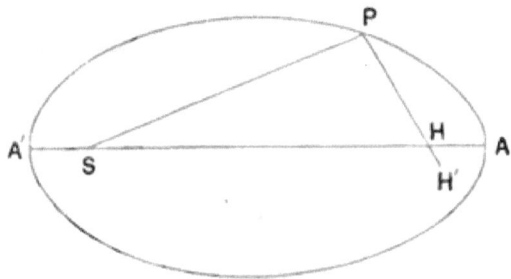

Let $APA'$ be the path of a particle moving about a centre of force **at** $S$ and let $H$ be the other focus.

When the particle **arrives** at $P$ let its velocity be changed from $v$ to $v+u$, the direction being unaltered; let $2a'$ **be the new major axis.**

Then we have

$$v^2 = \mu\left(\frac{2}{SP} - \frac{1}{a}\right); \quad (v+u)^2 = \mu\left(\frac{2}{SP} - \frac{1}{a'}\right)$$

∴ by subtraction, $\quad (v+u)^2 - v^2 = \mu\left(\frac{1}{a} - \frac{1}{a'}\right),$

giving the change in the major axis.

Since the direction of motion at $P$ **is unaltered the** new focus lies on $PH$; and, if $H'$ be its position, we have

$$H'H = (H'P + SP) - (HP + SP) = 2a' - 2a.$$

**Hence we have the** direction of the new major axis.

210. **If** the disturbing force be not tangential, the velocity it produces must be **compounded with** the velocity in the orbit to give the new velocity and direction of motion at the point $P$. This direction will be **the** new **tangent at** $P$; as in the last article we can now determine the

# 310 EFFECT OF A CHANGE IN ABSOLUTE ACCELERATION.

new major axis. Also from the fact that the moment of the velocity of the point about the focus is equal to $\sqrt{\mu \times \text{semi-latus-rectum}}$ we obtain the new eccentricity. Finally, by drawing a line making with the new tangent at $P$ an angle equal to that made by $SP$, and taking on it a point $H'$, such that $SP + H'P$ is equal to the new major axis, we obtain the new second focus, and hence the position of the major axis of the new orbit.

211. *Effect on the orbit of a change in the value of* $\mu$. When the particle is at a distance $r$ from the centre of force let the value of $\mu$ be changed to $\mu'$, and let the new values of the axis-major and eccentricity be $2A$ and $E$.

Since the velocity is instantaneously unaltered in magnitude we have

$$\mu\left(\frac{2}{r} - \frac{1}{a}\right) = \mu'\left(\frac{2}{r} - \frac{1}{A}\right),$$

which is an equation to determine $A$.

The moment of the velocity about $S$ being unaltered, $h$ remains the same.

$$\therefore \sqrt{\mu a (1 - e^2)} = \sqrt{\mu' A (1 - E^2)},$$

giving $E$.

The direction of the velocity at distance $r$ being unaltered we obtain the new position of the second focus, and hence the direction of the new major axis as in Art. 209.

212. *Motion in a straight line with an acceleration always directed toward a fixed point in the straight line and varying inversely as the square of the distance.*

The velocity when the point has described any distance from rest can be easily deduced from the results of Art. 205.

For the velocity at any point $P$ of the ellipse is given by

$$v^2 = \mu\left(\frac{2}{SP} - \frac{1}{a}\right).$$

Let $SA$ be $c$ so that $c = a(1 + e)$.

$$\therefore v^2 = \mu\left(\frac{2}{SP} - \frac{1 + e}{c}\right).$$

## MOTION IN A STRAIGHT LINE.

Let the minor axis of the ellipse be now indefinitely diminished, the major axis being unaltered in length; since $b^2 = a^2(1 - e^2)$, the eccentricity $e$ will become unity. Also the ellipse will degenerate into the straight line $SA$, and $P$ will become a point on $SA$.

The expression for the velocity will become

$$v^2 = \mu\left(\frac{2}{SP} - \frac{2}{c}\right),$$

where $SA = c$.

Hence, if a particle move in the straight line $AS$ under an acceleration $\dfrac{\mu}{(\text{distance})^2}$ towards $S$, its velocity when at a distance $x$ from $S$ will be given by

$$v^2 = 2\mu\left(\frac{1}{x} - \frac{1}{c}\right).$$

Also the time from $A$ to $S$

$= \mathrm{Lt}_{e=1}$ (Time of describing one half of the ellipse)

$= \mathrm{Lt}_{e=1}\left(\dfrac{\pi}{\sqrt{\mu}} a^{\frac{3}{2}}\right).$

$= \mathrm{Lt}_{e=1}\left\{\dfrac{\pi}{\sqrt{\mu}}\left(\dfrac{c}{1+e}\right)^{\frac{3}{2}}\right\} = \dfrac{\pi}{\sqrt{\mu}}\left(\dfrac{c}{2}\right)^{\frac{3}{2}}.$

**213.** *Motion of a projectile, taking into consideration variations of gravity.*

We have pointed out, **Art. 197**, that the attraction of the earth on a point outside the earth at **a distance $r$ from** its centre is $\frac{\mu}{r^2}$. Hence the motion of a projectile, the resistance of the air being neglected, is a particular case of Art. 205, one of the foci of the ellipse described by the particle being at the centre of the earth.

It will be noted that the value of gravity at the surface of the earth is $\frac{\mu}{R^2}$, where $R$ is the radius of the earth. Hence $\frac{\mu}{R^2} = g$, so that $\mu = gR^2$.

In the next article will be discussed a few elementary theorems on this portion of the subject.

**214.** Let $S$ be the centre of the earth and $P$ the point of projection. Let $K$ be the point vertically above

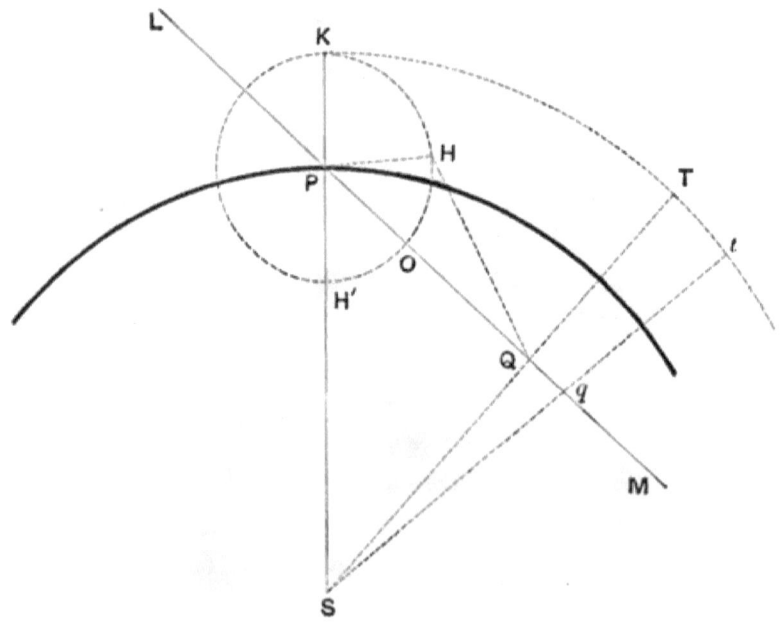

$P$ to which the velocity, $V$, of projection is due, so that, by Art. 212, we have

$$V^2 = 2\mu\left[\frac{1}{SP} - \frac{1}{SK}\right] = 2gR^2\left[\frac{1}{R} - \frac{1}{R+h}\right] \quad\ldots(1),$$

where $R$ is the **radius of** the earth and $PK$ is $h$.

If $H$ be the second focus of the path the semi-major axis is $\frac{1}{2}(R + PH)$. Hence, by Art. 205,

$$V^2 = \mu\left[\frac{2}{SP} - \frac{2}{R+PH}\right] = 2gR^2\left[\frac{1}{R} - \frac{1}{R+PH}\right].$$

By comparing this with equation (1) we have $PH = h$, so that the locus of the second focus is, **for a** constant velocity of projection, **a circle whose centre is** $P$ and radius $h$. It follows **that the** major **axis of the** path is $SP + PH$ or $SK$.

The ellipse, **whose foci are** $S$ **and** $H$, meets a plane $LPM$, passing through the point of projection, in a point $Q$ such that $SQ + QH = SK$. Hence, if $SQ$ meet in $T$ the circle whose centre **is** $S$ and radius $SK$, we have $QT = QH$. Since there is, in general, another point, $H'$, **on the circle of foci equidistant** with $H$ **from** $Q$ we have in general two paths for a given range.

The **greatest range on the plane** $LPM$ **is clearly** $Pq$ where $qt$ equals $qO$.

Hence $Sq + qP = Sq + qO + OP = Sq + qt + PK$
$$= SK + PK.$$

Therefore $q$ lies on an ellipse **whose foci** are the centre of the earth and **the** point of projection and which passes through $K$.

Hence we obtain the **maximum range.**

## MISCELLANEOUS EXAMPLES.

**1.** *Find the velocity acquired in falling through a height* h *to the earth's surface, taking into consideration variations in the acceleration due to gravity.*

If $R$ be the radius of the earth the acceleration at a point outside the earth is $\dfrac{\mu}{(\text{distance})^2}$, where $\dfrac{\mu}{R^2} = g$.

Hence the velocity of the particle on reaching the earth is given by

$$v^2 = 2\mu \left[\frac{1}{R} - \frac{1}{R+h}\right] = \frac{2gRh}{R+h}.$$

Hence the square of the velocity acquired is less than if gravity were constant by $2gh - \dfrac{2gRh}{R+h}$, that is, by $2gh^2/(R+h)$.

**2.** *If the earth were suddenly stopped in its orbit find the time that would elapse before it fell into the sun, neglecting the eccentricity of its orbit.*

Let $R$ be the radius of the orbit of the earth and $\omega$ its angular velocity about the sun.

The attraction of the sun is $\mu/(\text{distance})^2$, where $\dfrac{\mu}{R^2} = \omega^2 R$, so that $\mu = R^3 \omega^2$.

By Art. 212, the time required

$$= \frac{\pi}{\sqrt{\mu}} \left(\frac{R}{2}\right)^{\frac{3}{2}} = \frac{\pi}{2\sqrt{2}\omega} = \frac{\pi\sqrt{2}}{4\omega}.$$

But $\dfrac{2\pi}{\omega}$ = period of revolution of the earth = 365 days.

$\therefore$ time required $= \dfrac{365\sqrt{2}}{8}$ days = about $64\frac{1}{2}$ days.

If we do not neglect the eccentricity, $e$, of the orbit the time would lie between $64\frac{1}{2} \times (1 \pm e)^{\frac{3}{2}}$ days.

**3.** *A small meteor, of mass* m, *falls into the sun when the earth is at the end of the minor axis of its orbit; if* M *be the mass of the sun, find the changes in the major axis, in the position of the apse line, and in the periodic time of the earth.*

The new value of $\mu$ is $\mu \dfrac{M+m}{M}$. Hence, if $2A$ be the new major axis, we have

$$\mu\left(\dfrac{2}{a}-\dfrac{1}{a}\right)=\mu\dfrac{M+m}{M}\left(\dfrac{2}{a}-\dfrac{1}{A}\right),$$

so that
$$A=a\left(1-\dfrac{m}{M}\right)\quad\ldots\ldots\ldots\ldots\ldots\ldots\ldots\ldots\ldots(1).$$

If $S$ be the sun, $H$ the original second focus, and $H'$ the new second focus, we have $HH'=(SP+PH)-(SP+PH')=2a-2A$

$$=2a\dfrac{m}{M}.$$

$$\therefore\ \tan H'SH=\dfrac{HH'\sin H'HS}{SH-HH'\cos H'HS}=\dfrac{2a\dfrac{m}{M}\cdot\dfrac{b}{a}}{2ae-2a\dfrac{m}{M}e}=\dfrac{b}{ae}\dfrac{m}{M},$$

giving the angle through which **the major axis is turned**.

Also the new periodic time $=\dfrac{2\pi}{\sqrt{\mu\left(1+\dfrac{m}{M}\right)}}\cdot A^{\frac{3}{2}}$

$$=\dfrac{2\pi}{\sqrt{\mu}}\left(1-\dfrac{1}{2}\dfrac{m}{M}\right)a^{\frac{3}{2}}\left(1-\dfrac{3}{2}\dfrac{m}{M}\right)=\dfrac{2\pi}{\sqrt{\mu}}a^{\frac{3}{2}}\left(1-\dfrac{2m}{M}\right),$$

so that the periodic time is shortened by $\dfrac{2m}{M}$ of a year.

## EXAMPLES. CHAPTER IX.

1. A particle describes an ellipse under a force to its centre; shew that its angular velocity about a focus is inversely proportional to its distance from that focus.

2. A particle is describing an ellipse under a force to the centre; if $v$, $v_1$, $v_2$ be the velocities at the extremities of the latus rectum and major and minor axis respectively, prove that

$$v^2 v_2^2 = v_1^2(2v_2^2 - v_1^2).$$

# EXAMPLES ON CHAPTER IX.

3. A particle describes an ellipse about a centre of force in the focus; shew that the velocity at the mean distance from the centre is a mean proportional between the velocities at the ends of any diameter.

4. If a particle describe an ellipse about a centre of force in one of the foci, shew that its velocity at any point of its path may be resolved into two constant velocities respectively perpendicular to the major axis and the radius vector to the centre of force.

5. A particle describes an ellipse about a centre of force in the focus; shew that the angular velocity about the other focus varies inversely as the square of the normal at the point.

6. Shew that if the velocity of the earth at any point of its orbit were increased by about one-half it would describe a parabola about the sun as focus.

7. If a body be projected from the earth with a velocity exceeding 7 miles per second it will not return to the earth, and may even leave the solar system.

8. If the perihelion distance of the parabolic orbit of a comet be equal to the mean distance of the earth from the sun, find the number of days the comet will take from one end of its latus rectum to the other.

9. Given that there are 13 lunar months in a year, that the sun's angular radius as seen from the earth is 16', and that the earth's angular radius as seen from the moon is one degree, shew that the mean density of the sun is about one-third that of the earth.

10. Taking into consideration the variations in the value of gravity outside the earth's surface, shew that the maximum

range of a projectile on a horizontal plane passing through the point of projection is $4gu^2R^2/(4g^2R^2 - u^4)$, where $u$ is the velocity of projection and $R$ is the radius of the earth.

11. Shew that a gun at the sea-level can command $\dfrac{1}{n^2}$ of the earth's surface if the greatest height to which it can send the ball be $\dfrac{1}{n}$th of the earth's radius, and that the required direction of projection is inclined at an angle $\tan^{-1}\sqrt{\dfrac{n-1}{n+1}}$ to the horizontal, variations of gravity being taken into account.

12. If the velocity of projection from a point on the surface of the earth be such that directed vertically upward it would carry a body to a height equal to the earth's radius, prove that, if the direction of projection make an angle of $n°$ with the vertical, the range on the earth's surface will be $120n$ nautical miles.

13. A particle is describing an ellipse under a force to the focus, when it comes to the end of the minor axis the absolute force is diminished by one-third. Find the position and dimensions of the new orbit, and shew that the distance between its focus and its centre is bisected by the minor axis of the original orbit.

14. A particle is describing an ellipse under the action of a force to the focus, and when it arrives at the end of the minor axis the magnitude of the force is reduced to one-third of its original value; shew that the orbit will become hyperbolic and that its conjugate axis will be to the minor axis of the ellipse in the ratio of $\sqrt{3} : 1$.

15. When a particle is at the nearer apse of an ellipse described about the focus the absolute force is increased by a

small fraction, $\frac{1}{n}$th, of itself; when at the further apse the absolute force becomes less than the original value by this amount. Shew that the time taken in this revolution is equal to the original period reduced by $\dfrac{6e}{n(1-e^2)}$ of itself.

16. If when a particle is at any point $P$ of an elliptic orbit the centre of force be suddenly moved from the focus $S$ to the other focus $H$, prove that the intensity must be altered in the ratio $HP^2 : SP^2$ so that the particle may continue to describe the same path.

17. A body is revolving in an ellipse under the action of a force to the focus; on arriving at the end of the minor axis the law of force is changed to that of the direct distance, the magnitude of the force at that point remaining the same; shew that the periodic time will be unaltered and that the sum of the new axes is to their difference as the sum of the old axes to the distance between the foci.

18. If when a particle is at the extremity of the minor axis of an ellipse which it is describing about the focus, the centre of force be suddenly removed a small distance $aa$ toward the particle, the eccentricity will be unaltered but the major axis will be turned through an angle whose circular measure is $a\left[e^{-2}-1\right]^{\frac{1}{2}}$.

19. A planet, of mass $M$ and periodic time $T$, when at its greatest distance from the sun comes into collision with a meteor, of mass $m$, moving in the same orbit in the opposite direction with velocity $v$; shew that, if $\dfrac{m}{M}$ be small, the major axis of the planet's path is reduced by

$$\frac{4m}{M} \cdot \frac{vT}{\pi} \sqrt{\frac{1-e}{1+e}}.$$

# EXAMPLES ON CHAPTER IX.

**20.** A planet, when at the extremity of the latus-rectum of its path, meets with a shower of meteors which reduces its velocity by $\frac{1}{n}$-th without altering the direction; prove that the periodic time is reduced by $\frac{3}{n}\frac{1+e^2}{1-e^2}$ of itself.

**21.** The velocity of a body moving in an ellipse about the focus is increased in the ratio $1+n : 1$, where $n$ is small; shew that the corresponding increase in the eccentricity is $2n(e + \cos\theta)$ where $\theta$ is the angle $ASP$, $P$ being the position of the body, $S$ the focus, and $A$ the apse.

Shew also that the apse-line is turned through an angle $\frac{2n \sin\theta}{e}$.

**22.** A particle describing an ellipse about a centre of force in the focus is checked, when at $P$, by a small impulse $\gamma V^2$ along the tangent at $P$, where $V$ is the velocity at $P$; prove that the major axis is turned through an angle $\frac{2\gamma V}{e} \sin PSH$ and find the change in the eccentricity.

**23.** A particle is revolving in an ellipse about its focus and is at the end of the minor axis and moving away from the centre of force when it receives a blow which alters its path into a circle. Shew that the impulse of the blow is equal to $\sqrt{2\mu\frac{(a-b)}{a^2}}$ and find also the direction of the blow.

**24.** A body is describing an ellipse about a centre of force in the focus and when its radius vector is one-half the latus rectum it receives a blow which causes it to move toward the other focus with a momentum equal to that of the blow; find the position of the axis of the new orbit and shew that its eccentricity is $\frac{1-e^2}{2e}$, where $e$ is the eccentricity of the original orbit.

320   EXAMPLES ON CHAPTER IX.

25. A body is moving in an ellipse about a centre of force in the focus; when it arrives at $P$ the direction of motion is turned through a right angle, the speed being unaltered; shew that the body will describe an ellipse whose eccentricity varies as the distance of $P$ from the centre.

26. If a body (moving in an ellipse under a force to the focus) when at the mean distance have the direction of its velocity turned through an angle $a$, the eccentricity of the new orbit will be $e \cos a \pm \sqrt{1 - e^2} \sin a$, where $e$ is the original eccentricity.

If $a$ be small and at the same time the velocity be altered by $e \dfrac{\sqrt{\mu a}}{2b} a$, shew that the apse-line is unaltered.

27. A body, of mass $m$, describes an ellipse under a force to the focus whose absolute magnitude is $\mu$. When the body is at the end of the minor axis it receives a blow in the direction of the normal of magnitude $m\sqrt{\dfrac{\mu}{a}}$, the intensity of the force being at the same time doubled; shew that the magnitude of the axis of the new orbit is unaltered in magnitude, and that the new eccentricity is $\dfrac{1}{\sqrt{2}}(e + \sqrt{1 - e^2})$. If the intensity had remained the same, shew that the new orbit would have been a parabola.

28. A body is moving in an ellipse about a force in the focus; when at the end of the latus rectum through the centre of force it receives a blow which makes it move at right angles to its former direction; shew that the new orbit will be a hyperbola concentric with the ellipse if the ratio of the velocities before and after the blow be $e : 1$.

# ANSWERS TO THE EXAMPLES.

## CHAPTER I. (Pages 50—55.)

1. $\dfrac{3h}{8}$.

2. $\sqrt{gh}$; zero and $\sqrt{gh}$, where $h$ is the height of the plane.

3. $\dfrac{16000\pi}{33}$.

5. $\sqrt{u^2 + fs\left(1 - \dfrac{1}{n}\right)}$.

9. $\tfrac{2\,2}{3\,3}$ ft.-sec. units; $\tfrac{4\,4}{1\,5}$ ft.-sec. units; 40 miles per hour.

10. 60 miles per hour; 44 secs.; 1936 **feet**; 8 secs.

14. If $P$ be the point of the circle furthest from the given straight line, and $Q$ be the point moving in the circle, the velocity of $Q$ with respect to the other is $PQ \cdot \omega$ perpendicular to $QP$.

20. $\omega\sqrt{\dfrac{2h^3}{g}}$, where $\omega$ is the angular velocity of the earth about its axis.

25. (1) The chord makes an angle of 60° with the vertical; (2) the chord makes an angle $\sec^{-1}\sqrt{3}$ with the vertical.

27. Through the end of the latus rectum and inclined at 60° to the horizon respectively.

29. The length of the chord is $b\sqrt{2}$.

37. A circle passing through the lowest point of the given circle and touching it there.

L. D.

# ANSWERS TO THE EXAMPLES.

## CHAPTER II. (Pages 103—113.)

**1.** $\frac{g}{2}, \frac{g}{3}$, and $\frac{5g}{12}$ poundals respectively.

**4.** Weight of $9\frac{200}{324}$ tons.    **6.** 54,650,956,800 absolute units.

**7.** $20\sqrt{2}$ feet per second; 560,000 foot-pounds.

**8.** $68\frac{68}{135}$ H.-P.

**10.** Weight of $11\frac{3}{8}$ tons; $28,233,333\frac{1}{3}$ foot-pounds.

**11.** About 50·4 and 23·1 feet per second respectively.

**12.** ·938 tons weight.    **13.** $1260\frac{5}{12}$ poundals.

**14.** $180\frac{444}{1000}$.    **15.** 7270.    **16.** $11\frac{3}{14}$ tons.

**17.** 1 minute $42\frac{2}{3}$ seconds.

**19.** The parliamentary train by 9·46 per cent.

**21.** $9\frac{9}{11}$ miles per hour.

**23.** The straight line which bisects the distances between the eaves and the ridge and ground respectively.

**24.** $\frac{8}{5}\sqrt{6}$ feet per second; $\frac{3}{5}\sqrt{30}$ feet per second.

**27.** $8mm'/(m+m')$ units of impulse; $8m/(m+m')$ feet per second.

**29.** After $M$ reaches the ground, $M'$ describes a distance $h\frac{M-M'}{M+M'}$ on the plane and then returns.

**30.** 660,000 foot-lbs.; 30 H.-P.    **32.** $\cdot 16\dot{9} : 1$.

**36.** ·88 foot-tons.    **41.** 200 feet above; 1000 feet below.

**42.** $50g\left\{\frac{m_1-m_2}{m_1+m_2}+\frac{3(m_1-m_2)-m_3}{m_1+m_2+m_3}\right\}$.

**43.** $\frac{u}{g}$, where $u$ is the common velocity.

**45.** $\frac{1}{4}\frac{n+1}{n}ft^2$, where $f$ is the acceleration produced by the force $\frac{1}{4}ft^2$.

**52.** $(P-W)g/W$; the tension at any point distant $x$ from the lowest point is $\frac{x}{a}P$, where $a$ is the whole length of the chain.

ANSWERS TO THE EXAMPLES.

## CHAPTER III. (Pages 129—142.)

**1.** The acceleration becomes $(M - 8m + 7\mu)g/(M + 64m + 21\mu)$.

**6.** $\dfrac{m_1 - 2m_2}{m_1 + 4m_2}g$; $2\dfrac{m_1 - 2m_2}{m_1 + 4m_2}g$; $\dfrac{3m_1 m_2}{m_1 + 4m_2}g$ poundals; it is assumed that all the strings are vertical.

**9.** If $M$ be fixed its acceleration is $(M - m - m')g/(M + m + m')$.

**12.** $4mm'/(3m - m')$.

**14.** $W_1$ will remain at rest if $\mu W_1 = W_2 + 4W_3 W_4/(W_3 + W_4)$.

**17.** The velocity of the weight struck is five times that of each of the others and is in the opposite direction.

**19.** 1005·2 feet per second.

**20.** About 3580 and 3100 feet per second respectively.

**25.** The ratio is $M : M + m$, where $M$ is the mass of the bucket and $m$ is the mass of the frog.

**26.** $\dfrac{P}{m}\dfrac{m - m'}{m + m'}$.

**27.** $\dfrac{u}{2}\dfrac{a(a+b)}{a^2 + ab + b^2}$; $\dfrac{u}{2}\dfrac{b(a+b)}{a^2 + ab + b^2}$.

**28.** $\dfrac{mu^2}{3 + \cot^2 a}$ where $m$ is the mass of each particle, $u$ the velocity with which $C$ is projected, and $a$ the angle $ABC$.

**30.** $57\tfrac{1}{2}$ H.-P.    **31.** $2\tfrac{8}{11}$ H.-P.

**32.** $125\pi/96$ revolutions; $125\pi/24$ revolutions.

**34.** The ring comes to rest when its length is $2\pi x$ where
$$W'\left[\sqrt{r^2 - a^2} - \sqrt{r^2 - x^2}\right]a = \pi\lambda(x - a)^2,$$
$W'$ being the weight of the ring, $r$ the radius of the sphere, $2\pi a$ the initial length and $\lambda$ the coefficient of elasticity of the ring.

## CHAPTER IV. (Pages 164—166.)

**1.** $ga^2$ feet; $a$ seconds; $\dfrac{b}{a}$ lbs.

**2.** 4·3...grammes; 18·21...metres; 5·45 seconds.

**3.** 2400 feet; 15 seconds; 600 lbs.

**4.** 8800 yards; 5 minutes; $54\tfrac{6}{11}$ tons.

**5.** 1 mile; 8 minutes; $99\tfrac{17}{77}$ tons.

**6.** $3:1$; $9:1$; $2:15$.    **7.** $[T]^2[M][E]^{-1}$.

**8.** $\sqrt{\dfrac{E}{gR^2}}$ = about 31300 lbs.

**10.** About $33 \times 10^7 \pi$ lbs.; about $3{\cdot}35 \times 10^{-12}$ lbs. weight.

**11.** $[L]^3[T]^{-2}[M]^{-1}$. **12.** $6{\cdot}488 \times 10^{-4}$ dynes.

**13.** $3{\cdot}98 \times 10^{20}$.

### CHAPTER V. (Pages 191—201.)

**6.** A circle of about 91 miles radius. **8.** $2\tfrac{1}{4}$ lbs.

**9.** $\dfrac{(u^2+ga)^2}{2gu^2}$, where $u$ is the velocity of the centre of the wheel and $a$ is its radius.

**12.** Its intersection with the horizontal moves in the same manner as a point which starts with velocity $\sqrt{3}u$ and moves with acceleration $-\sqrt{3}g$.

**30.** In each case the maximum range will be found by drawing the enveloping parabola; the angles of projection are $\tan^{-1}\sqrt{\dfrac{h}{h+k}}$ and $\tan^{-1}\sqrt{\dfrac{h}{h-k}}$ respectively, where $k$ is the height of the cliff and $h$ the height to which the velocity of projection is due.

**34.** Let $A$ and $B$ be the two given points, $B$ being the lower; draw $BK$ vertical and $AB$ horizontal to meet it in $K$; the focus of the required path is a point, $S$, on $AB$ such that $2AS = AB - BK$; the required point, $C$, on the horizontal plane is such that $2CS \cdot AB = AK^2$.

**36.** $6{\cdot}1$ inches. **39.** $1{\cdot}92$ lbs.

**42.** The latus rectum is unaltered; also, if $S$, $S'$ be the old and new foci and $P$ be the point of contact, then $S$, $S'$, and $P$ are in a straight line, and $S'P : SP :: (m-m')^2 : (m+m')^2$, where $m$ and $m'$ are the masses of the particles.

### CHAPTER VI. (Pages 227—240.)

**6.** The vertex of the new path is in the horizontal line through the starting point at a distance $\dfrac{3a}{2}$; the focus is at a depth $\dfrac{a}{8}$.

**7.** $2\tfrac{7}{8}$.

**22.** The condition is that the equation $1 - 2e^n + e^{2n+r} = 0$ should give an integral value of $n$ for an integral value of $r$.

## ANSWERS TO THE EXAMPLES.   325

**42.** The velocity of each is ultimately $mv/(M+m)$.

**43.** $12\frac{1}{4}$ feet; if the height be less than this, and the train be long enough, the ball will rebound from a point of the train behind the point that it first hits.

**44.** If $\tan a$ be $> \dfrac{1-e}{1+e} \cot \lambda$ the horizontal velocity will be destroyed before **the vertical velocity** and the particle will rebound vertically.

## CHAPTER VII. (Pages 268—276.)

**4.** 30,000 poundals.

**5.** $\left(\dfrac{\lambda}{r} - \dfrac{M\omega^2}{2\pi}\right)$ poundals **per unit of length,** $\lambda$ being the coefficient of elasticity, $M$ the mass of the band and $r$ its radius.

**7.** 24·31 hours roughly, taking $g = 32$ and $\pi = \frac{22}{7}$.

**11.** $\sqrt{ag} \sec a \sqrt{3 + 2\sin a - 4\sin^2 a}$.

**12.** Its mass must be diminished in the ratio **1 : 4**.

**14.** $\sqrt{gl}(\sqrt{2-1})$; $1 : \sqrt{2}$; $\dfrac{\pi}{2}\sqrt{2\sqrt{2-2}}$.

**26.** $r \operatorname{cosec} \dfrac{\theta}{2} = \text{constant}$.

**27.** The lesser ball remains at the bottom of the tube and the **greater** rises through **a** height equal to one-half the radius of the tube.

**30.** $2\cos^{-1}\dfrac{g}{l\omega^2}$; with a load $M'$ the angle is $2\cos^{-1}\dfrac{M+M'}{M}\dfrac{g}{l\omega^2}$, where $l$ is the length of a rod and $\omega$ the angular velocity of the system.

**36.** $18\pi a/u$ seconds; $mu^2/a$ poundals; $mu^2/6a$ poundals.

**37.** $2\sqrt{7ga}$.

**39.** If $\rho$ be $> 4$ the particle $A$ **will always be at rest.**

## CHAPTER VIII. (Pages 295—299.)

[$\pi$ has been taken to be $\frac{22}{7}$ and $g$ to be **32**.]

**1.** 165 complete oscillations nearly.  **3.** 32·15.
**4.** 1 : ·999953.   **5.** 1·0046 : 1; the depth is 1·85 miles.
**6.** 10·8 seconds; ·0097 inch.   **8.** 432.   **9.** 981.

326   ANSWERS TO THE EXAMPLES.

**10.** ·0081 inch.    **18.** $\left(1 - \dfrac{1}{4n}\right)$ sec.

**25.** A cycloid; a prolate cycloid.    **26.** 45°.

**29.** The equations to the paths are

$$\cos^{-1}\frac{x}{a} - 3\cos^{-1}\frac{y}{a} = 0, \quad \text{and} \quad \cos^{-1}\frac{x}{a} + 3\sin^{-1}\frac{y}{a} = 0,$$

where $a$ is the common amplitude.

## CHAPTER IX.  (Pages 315—320.)

**4.** The component velocities are $\dfrac{h}{b^2}ae$ and $\dfrac{h}{b^2}a$ respectively.

**8.** 219 days nearly.

**13.** The new major axis is $4a$; the new eccentricity is $\tfrac{1}{2}\sqrt{1+3e^2}$; the new major axis makes an angle $\tan^{-1}\dfrac{b}{2ae}$ with the original major axis.

**22.** The increase in the eccentricity is $\dfrac{1-e^2}{e}\gamma V\left(1 - \dfrac{V^2 a}{\mu}\right)$.

**23.** The direction of the blow bisects the angle between the minor axis and the radius vector to the centre of force.

**24.** The axis is unaltered in direction.

www.ingramcontent.com/pod-product-compliance
Lightning Source LLC
Chambersburg PA
CBHW030004240426
43672CB00007B/819